SHROUD

BY ADRIAN TCHAIKOVSKY

Shadows of the Apt series
Empire in Black and Gold
Dragonfly Falling
Blood of the Mantis
Salute the Dark
The Scarab Path
The Sea Watch
Heirs of the Blade
The Air War
War Master's Gate
Seal of the Worm

Echoes of the Fall trilogy
The Tiger and the Wolf
The Bear and the Serpent
The Hyena and the Hawk

The Final Architecture trilogy
Shards of Earth
Eyes of the Void
Lords of Uncreation

Guns of the Dawn

Children of Time
Children of Ruin
Children of Memory

The Doors of Eden

Alien Clay

Service Model

Shroud

ADRIAN TCHAIKOVSKY

SHROUD

TOR

First published 2025 by Tor
an imprint of Pan Macmillan
The Smithson, 6 Briset Street, London EC1M 5NR
EU representative: Macmillan Publishers Ireland Ltd, 1st Floor,
The Liffey Trust Centre, 117–126 Sheriff Street Upper,
Dublin 1, D01 YC43
Associated companies throughout the world
www.panmacmillan.com

ISBN 978-1-0350-1379-1 HB
ISBN 978-1-0350-1380-7 TPB

1 3 5 7 9 8 6 4 2

A CIP catalogue record for this book is available from the British Library.

Typeset in Dante MT by Palimpsest Book Production Limited, Falkirk, Stirlingshire
Printed and bound by CPI Group (UK) Ltd, Croydon, CR0 4YY

FSC www.fsc.org
MIX
Paper | Supporting
responsible forestry
FSC® C116313

Dedicated to the memory of Roland Oughton: friend, adviser, gamer, swordsman and strategist. Also a damn fine advance reader. Forever to be missed.

ACKNOWLEDGEMENTS

Thanks to William Bains for all your time and scientific input while brainstorming this one. Further thanks to Michael Czajkowski and Dermot Dobson for additional scientific advice and assistance. Thanks also to Mike Carey, Roland Oughton and Antony Johnston for some fresh eyes*.

* On the manuscript. Not actual eyes.

SELECTED CREW OF THE *GARVENEER* (IN ORDER OF SENIORITY)

Sharles Advent – Opportunities, Chief Director
Terwhin Umbar – Opportunities, Technical Oversight
Jadi Timorai – Safeguarding, Director
Oswerry Bartokh – Special Projects, Director
Hari Skien – Special Projects, Exogeologist
Shemp Rastomaier – Special Projects, Exobiologist
Juna Ceelander – Special Projects, Administrator
Mai Ste Etienne – Special Projects, Macro Engineer
Mikhail Jerennian – Special Projects, Data Engineer
Ducas FenJuan – Remedial Work, Exobiologist

PROLOGUE

The couch's design had been cribbed from those on the *Garveneer*, which had cushioned everyone's sleeping forms while the ship accelerated through the interstellar void, so it was considerate enough to wake me up after impact. Conscientious as a servant. Or a doctor. Waking me into pain and darkness. The sour residuum of all the terror and panic I'd been caught in the throes of before I blacked out.

Everything hurt. Out of the chorus, specific pains spoke up. I was a mess of discrete bruises, like I'd gone three rounds with a pugilism automatic. But beyond that, everything hurt me. The world pressed on me with such universal force I wondered, *Are we travelling? Are we still accelerating?* As though I was still in the ship.

Fighting against the gelatinous suction of the couch. Clawing an arm free of it, groping for the controls which should have been beside me on the *Garveneer*. My fingers met only cold panels. Dead, unseen, existing merely in the contact of my fingers with their smooth surfaces.

I remembered: I wasn't on the ship. The oppressive weight of my own body, which was half suffocating me, wasn't just the artefact of man-made motion. And the thrum of directionless sound wasn't the thunder of distant engines communicated to me through the *Garveneer*'s superstructure.

I heard the whimpering of my laboured breath, loud in my ears.

No. Please. We didn't.

But we had. We hadn't meant to, but events had overtaken us.

Is it just me? Am I alone?

The rush of my panicky, straining lungs would have eclipsed an army.

We fell.

From grace. From orbit.

Dim light rose on all sides. Something lived beside I alone. The feeble motion of my hand had woken it. Faint illumination seeped out into the air, touching across the interior of the spherical space around me, yet some panels remained dark. I wasn't the only one who'd taken a battering.

'Let me see.' Getting the words out was an effort. A wasted one. It wasn't an instruction any listening system could have helped with. I mustered some jangled thoughts. Recited the mantra: *Juna Ceelander, crew of the* Garveneer. *Special Projects team. Administrator.* Administrator meaning problem-solver and general factotum. *Not supposed to be here. Not in this thing. Not down here on . . .*

Shroud.

'External cameras on.' Probably still not the precise wording, but if there were systems capable of listening then they'd interpret what I wanted and give me something. Anything.

I waited, ears full of myself straining to live, and the echoes of my voice.

Panels darkened, converting themselves into screens for my better edification and enlightenment. I stared into them, waiting to see what was out there.

And it was nothing.

The cameras were defective, I assumed. Or else the screens were defective.

But it wasn't them. It was the world. The lightless, crushing, freezing abyss that was the world of Shroud.

I started to whimper again, then forced myself to stop. *It could be worse. There's still hope. It could be worse.*

Distantly, through all the multi-layered and vastly over-engineered hull, there was a dull metal sound. All the many tons of my surrounding structure shifted slightly, as through the idle exertion of an immense strength. I heard a slow, indolent scrape, as if some hard claw was being dragged down the outside of the pod. Out there in the murderous, clenching dark.

I stared at the screens, still seeing nothing at all, and tried not to scream.

PART ONE
ABOVE / BELOW

1.1 LIGHT

The world roared.

It was a moon, in truth. A tidally locked orb, slaved to the gravity of a roiling gas giant, out towards the waist of the star system Prospector413 (annotation: approved for exploitation). This system of cold dead worlds and mineral wealth was about to form part of the Third Stage Commercial Expansion out of Earth. The net that humans were throwing out into the universe, to find somewhere to harvest and build and live. If you could call it living.

We hadn't even been supposed to end up in orbit here. I remember waking up with the rest of Special Projects, dragged out of the near-freeze hibernation they had us all in. Biologically barely changed and yet somehow feeling incalculably older and more worn out. Each of us sloughed out of our couches like our own cast-off skins. Bartokh, our boss and very specifically my boss, clapped his hands together, striding out naked on his spindly legs, slapping at his gut because he was on enough of a worth-wage to acquire one. He'd been up for an hour already, getting over the cosmic hangover stage of proceedings without ever bothering to throw a scrap of clothing on. Now he was reminding us all of the abrasively cheerful leadership style he practised, which felt like sandpaper on the frontal lobes.

'What?' came the demand from Hari Skien, the geologist. 'What, for the love of God? Someone give us some perspective. Is it aliens?'

'Of course it's not aliens,' said Shemp Rastomaier, the team biologist, who'd basically given up on ever being able to do his job. By now the rest of us were moaning at the pair of them to either shut up or screw, because they'd been at each other's throats when we went under too. Apparently hibernation had enacted some kind of saved-state on their brains and they were right back at it now we were awake again.

Bartokh, the old, the gleeful, cackled and said, yes, maybe this time it actually was aliens.

Nobody believed him.

But it was. This time, it was.

The *Garveneer* Composite Mission Vessel had entered the Prospector413 system with a plan to start scouring the patchy asteroid belt and some of the outer planets first off. For this mundane purpose they'd wake the bigwigs, who comprised the select management team known only as Opportunities, and then probably Acquisition Teams One through to Seven or something. Except they hadn't even started setting up before they caught the signals. From a moon of a gas giant they'd have got around to inspecting eventually, except it was demanding attention right now.

It roared. Howled into space, shrieking its pain far out across the void in a thousand tongues. A constant, all-frequencies storm of radio traffic that blazed bright on every instrument we had.

This was, Opportunities decided, exactly why it needed

the boffins of the Special Projects team around. Because something weird was happening, and spectrographic analysis of the moon's atmosphere made for some interesting reading, so maybe there was some really useful stuff to be harvested over there. They took us off the shelf where we'd been stored, and designated a detachable module as Special Projects territory. They then split us off from the main mission's planned approach, and spat us out of the *Garveneer* to visit this screaming prodigy and see what all the fuss was about.

When we arrived, we found only darkness.

A well of night, blazing with electromagnetic activity. So much signal that it became only noise. Of the actual moon itself, we could decipher nothing. Our instrumentation might represent a constant running battle between engineer and entropy, but it was still the best that human ingenuity had designed. It couldn't hear itself over the tumult, and nobody down there was listening. Every electronic enquiry of ours vanished into the storm without leaving a forwarding address or a scrap of useful data. Our every hello was abraded away into nothing by the ceaseless shouting of the world below. A fleet of drones was able to skim the surface and return with some basic info about the upper atmosphere, but deeper enquiry just vanished into that cloak of noise. The six of us in Special Projects ended up split into at least eight factions, arguing over what could be causing such irregular, constant, chaotic transmissions. All of us dancing about the question of *Life, could it be life?* A whole civilization with their radios turned up to top volume, down there in the dark?

* * *

Eventually, something came back. *Eventually* meaning weeks of us measuring and experimenting, designing and fabricating, failing, revising. Scrounging for scraps of data around the edges, scavenging the low-hanging fruit of the moon's high-atmosphere composition. We drifted there, suspended over the roaring world, until even our regular human senses started to feel the moon's insane, hostile bellowing. Like being in a cell next to a crazed berserker. Bad dreams, ill tempers, the failure of attempt after attempt only worsening matters. Everything we dropped into that void was crushed, destroyed, malfunctioned, crashed, or just vanished into utter silence. We felt like we were battering ourselves against the moon's intractability like a fly at a window. Until, at last, something came back.

Deep Drone Fourteen, the previous thirteen having been lost to the impenetrable night below.

What returned of the drone was its very heart, torn from its body. Because the body was heavy. Because heading downwards was easy, but the moon was larger and greedier than Earth, so you couldn't just come back the regular way. The one thing it definitively wasn't, given the atmospheric composition, was rocket science.

This telltale heart was a heavily shielded datastore, the receptacle for all the instruments that had been left behind with the rest of Drone Fourteen's casing. Even getting *that* back up had proved powerfully challenging. We'd all been in the dark, literally, waiting to see if finally, this time, something would escape the screaming oblivion that was down there.

And it had. Which was how I, Juna Ceelander, had a first glimpse of life on Shroud.

* * *

Static, mostly, in that first watching, before Jerennian cleaned it up. Static because the constant chorus was far louder down there and, despite our best attempts, had crept into the data. Abrasive as a digital sandstorm, chewing at the edges of everything the drone had tried to preserve. For a moment the six of us sat there and wanted to slit our collective wrists in despair. Opportunities, the *Garveneer*'s high command, had been riding Bartokh hard for results. Which meant Bartokh had been riding us, and we all knew that if we kept on drawing a blank on usable data, they'd put us back on the shelf in the hibernation bay and maybe never wake us up again. You had to earn your keep in the commercial exploitation fleet of the Concerns, and all we'd been doing was guzzling resources.

Then, a glimpse: a dozen seconds of recorded video that was more than just all-direction snow and a vague suggestion of form. The electronic eyes had been left behind with the rest of the machine on the planet's surface, but the images they'd seen had come back to us in orbit, through all our improvised measures.

It was dark, but we'd known it would be. The drone had the most powerful lamps we could fit it with, which didn't cut far into the murk down there, but in those few seconds of imagery it was enough for us.

Not the barren moonscape we'd all been privately expecting, because so many failures, after so long, bred only pessimism. But it wasn't rocks, it was *things*.

A knotted tangle of barbed wire. A snarl of briars, save that the drone was twice the size of a human being and the briars still reached high overhead in an arcing tangle. The cameras angled downwards saw more of the same, interlaced like basketwork, riddled with irregular gaps that led beyond the light's reach. They shifted against one another as though the ground was breathing. The arching, knotting strands above showed a fractured punctuation, which was where the drone had torn through. Their edges were jagged, hollow, honeycombed.

The briars were also segmented, and they bloomed. Dimly glimpsed florets spread and waved, like the fans of tubeworms. With the wind? No, each was in motion separately. Beyond the shattered ceiling, lines extended upwards into the staticky blackness; taut, shifting, like the strings of kites.

Something moved at the edge of the light, intruding straight into it, with no sense of it registering the illumination. We glimpsed a body made of intricate latticework, a shift of motion within, seeming in that moment more like a child's clockwork toy than a living thing. The way it moved was profoundly unnatural to the human eye, though in those brief moments the reason wasn't clear. Something about its balance and gait. It was bigger than the drone, and its twisting progress was a slow amble. Its exterior shivered and shimmered, the light gleaming rainbows from a mist of fine hairs surrounding it. Then, as though in response, the tides of static rose and drowned the images. Shroud was keeping its secrets.

There were four hours of recordings locked inside the drone's retrieved data core. Of the visual records, those twelve seconds were all that were readily interpretable at first glance. But it was enough to let us know that every trial we'd gone through to reach this point had been worth it. Enough for us to show to Opportunities, to justify our continued waking existence.

Beyond the visual data, we retrieved a wealth of physical

test results, revealing a great deal about what it was like down on the surface of Shroud. Which was to say pretty damn unpleasant, for a human. Nobody was going for a stroll down there any time soon. Or ever. If you'd asked me right then, I'd have told you that was one thing everyone was certain of.

I mean, you'd have to be *mad*.

Although I suspect Bartokh was having ideas even then. I think he knew how it was going to go.

Bartokh gave us our marching orders straight away. Jerennian was to get to work trying to salvage more from the drone recordings. Everyone else was to work on developing better drones. Bartokh wanted more: atmosphere samples from ground level, specimens. He wanted something here in orbit that he could see, put under a microscope, then exhibit to Chief Director Advent when the big man came calling. But Shroud was jealous, and we would have to fight it every step of the way.

Three months and fifty-nine drones later, Shroud would win.

1.2 LIGHT

They called it the Second Bottleneck.

The first bottleneck – which for some reason didn't merit the capital letters in the history books – had been on Earth. This was the great climate and resource crisis that wrecked so much, and came so close to ending everything humanity had built.

But we scraped through it, and those same books tended towards a tinny, slightly strained paean of manifest destiny, and how all that suffering and desperation had obviously been necessary, because otherwise we wouldn't have ventured out into space and achieved *this*, and maybe don't think too hard about it, how about that?

The social structures which survived the first bottleneck and led us into the second – the Second, with capitals – were the products of that desperation. Incorporated Fealty. This was what we learned, in school. How 'Megasocial Opportunity-Exploitation Concerns' had been the salvation of our species. What they used to call corporations, except that was back when the presence of other ways of being gave the term some meaning. The rhetoric in the textbooks affirmed that only the big corporations could have done it. Unfettered as they were by the bonds of accountability and reciprocal responsibility, and possessed of foresight and tools which permitted a superhuman prognostication. The whole party line, you know how it goes. Feel free to sing along.

We saw only footnotes about all the other stuff that went on. The Inter-Concern Wars which devastated so many in-system colonization attempts. The way that, even as we were trying to build out from Earth, we were trying to tear ourselves back down too. All the stupid decisions which created the very last narrowing of that first bottleneck. Because despite all that, the fanfare and flag-waving part of the business was true, too.

The people who survived did so with the blessing of the Concerns, and only with that blessing bestowed in return for a pledge of service, as every child out of the habitat tanks can tell you in obedient chorus every morning. This was how we lived. And even if it was crammed together in the great orbital arcologies of the tanks, we *did* live, so we can all grumble as much as we want, but who are we to say there was another way? The six of us in Special Projects had given our lives to the *Garveneer* project, sure, but we were only around to do so because our progenitors had given their own lives to other Concerns, and thus been provided for, and bred, and given the universe *us*.

We were all properly grateful, I assure you.

We were lucky enough to live into the Second Bottleneck as I say, and humanity's attempts to escape it. This time the Concerns were driven to look outwards at the wider galaxy. Long-term thinking. Expansion. By that point the solar system itself was practically an automated factory. Every part of it that hadn't been blown up, or ferociously irradiated beyond even industrial use, was being industriously stripped. We could have stuck around, it was true, clinging to our mother system's apron strings. We'd have lasted a while. But the driving ambition of the Concerns was always expansion and growth. The avoidance of stagnation. The solar system was cramped now, mostly because a lot of the potentially usable parts of it hadn't really survived the internecine devastation at the end of the first bottleneck. In order to survive the Second Bottleneck, the Concerns needed to look further afield. So rather than a future penned up in a single, vulnerable planetary system, shackled to the dead poison cinder of Earth, we were heading outwards.

The *Garveneer* was just one of a hundred expeditions sent to every exoplanet-bearing system that looked hopeful. Our mission: to hunt out opportunities. To harvest resources. To create self-sufficient, resource-generating waystations that would serve to resupply the next wave of human exploration. The idea was that, in the future, the galaxy would become a network of stations to everywhere, settlements and extraction operations and shipyards, passing humanity hand over hand from one star to the next, and the next. The Concerns, perpetual economic powerhouses that they were, thought past the death of stars and the exhaustion of planets towards the immortality of their particular model of human society. We were going to spread ourselves across the near-side section of the galaxy like a rash.

Bartokh liked his face-to-face meetings. We could have done everything more efficiently from our workstations, but Bartokh was old. He had been in and out of the hibernation couches more than anyone. Once I realized this, I began to appreciate his insistence that we all actually dragged ourselves into the same room every so often. Because in the rush to hit targets and quotas, it was easy to forget you were human.

Needless to say, I prepared the presentation. Modelling things; that was part of my skillset. Not exactly the most useful thing to take into the field, but then I never thought I'd end up going *into* the field. I was just Director Bartokh's administrative assistant.

Despite the generally uninformative nature of Shroud – our unofficial name for the moon, but one that stuck – we knew some things from our initial flyby, and from Drone Fourteen. Its distance from the star (via its planet, Prospector413b, on average 498 million km). Size (thirty per cent larger than Earth); orbit (tidally locked, 112 hours to circle the waist of a giant larger than Jupiter); gravity (1.8 that of Earth, partly from its size, partly what was probably an unusually massive iron core); atmosphere (anoxic, volatile, thick as soup; basically majority nitrogen but with dangerously high levels of free hydrogen, plus a whole lot of ammonia, methane and other more complex stuff); pressure at wherever Fourteen had fetched up (twenty times Earth at sea level); temperature at the same point (minus thirty-five centigrade).

Those were the drawbacks, really. All the reasons you'd never want to go there. The pluses were that minus thirty-five was actually really balmy for a moon of a gas giant this far out, which meant that a combination of planetary radiation, vulcanism and a tiny greenhouse effect were obviously doing their best under trying conditions. There was also a lot of stuff in the atmosphere we couldn't readily explain, like why the high levels of hydrogen hadn't all bled off into space, or where the ammonia was coming from, or a variety of other puzzlers which Rastomaier had been scratching his head over. Without that, it wasn't anywhere we'd have been looking for life, save for the fact that it was literally yelling at us to come look.

Rastomaier was a miserable streak of a man, with a droopy moustache and a combover. Short and skinny, like most of us who'd come out of the orbital arcologies of the habitat tanks, and because nobody on Special Projects had the wage-worth for big dinners, except Bartokh. Before Fourteen's little light-show, Rastomaier had only been talking of life you could fit under a microscope slide. Skien, his opposite number on the rocks front, was shorter, balder, and their sourness was of a more adversarial flavour. Denying the whole life hypothesis, they'd been positing some way you could torture the moon's geology to make it scream like it was doing, and maybe weep ammonia tears too. Nobody had actually expected Big Life, though. Nobody had been looking for a son of a bitch bigger than the actual drone. It wasn't going to fit in a sampling canister, that was for sure. It was a whole macrobiotic ecosystem down there – life on an Earth scale, a dinosaur scale. We'd found Shroud during an age of alien giants.

I'd modelled the planet based on what we knew. Meaning pretty damn little, and we couldn't even formulate much of an idea of topography because, again, almost everything we sent down there just vanished. Or maybe was eaten, given what Fourteen had seen. I'd tried to model the creature from the clear footage, too, but failed. Too fleeting, too alien. Nothing I could come up with actually fitted the data. So that was just one other part of my role I couldn't do.

My job description was Bartokh's assistant. Having an actual human being serving as his assistant was the main privilege of his rank. What my job *actually* entailed was Bartokh knowing he could reliably unload most of his unwanted chores onto me, and I'd get them done in a close-enough approximation of his working style, so he could then show them to Chief Director Advent in Opportunities and make himself look good. My job was also serving as unofficial liaison between Bartokh and everyone else, as well as between Skien and Rastomaier, and between the two engineers and the rest of the team. Basically, I was perfectly positioned for everyone to blame whenever anything went wrong.

Due to this unspoken arrangement, what my *skillset* included was an aptitude for being the go-between amongst any two given factions in the team. A major in non-confrontational communications, meaning that whichever combination of the crew were currently pissed off at each other, they could sit down with me, and I would talk them through whatever the current spat was and smooth it all over. Honestly, I don't like to blow my own trumpet, but when you're six people stuck in a can and under the hammer of Opportunities to come up with results, I reckon my role was as important as any of the science.

The *Garveneer* Special Projects module was a configurable feast of a vessel. We hit orbit and the whole show turned itself from a ship with a front and a back, to a station hanging over the dark orb of the moon, which itself was just a jet bauble circling the enormous, purple-red-blue storm of the planet Prospector413b. *Garveneer SP* in its orbital identity always reminded me of a bug. One of those really fat bugs, whose wings don't look big enough to keep it in the air. Of course the thing which kept us up was our trajectory around

Shroud, fast enough to stop us falling into its hungry gravity, but not fast enough to just send us zipping off into space. Our wings were our solar arrays – multi-layered, hyper-efficient and still struggling to draw much from Prospector413, the system's star. The bulging belly of *GSP*, directed towards the moon, was hollowed out to make space for our laboratory-workshop. That was where we diligently gathered information about Shroud, so we could make a profitable use-case scenario to Opportunities and justify our ongoing existence and resource budget. The back of the ship, facing away from the moon and sandwiched between workspace and solar panels, was a patchwork of cramped little bunk rooms, plus our social space – the circular room where almost everyone came together to bitch about whoever wasn't present at the time. There was no gravity throughout any of it, but we were more than used to this. A growing proportion of the human species lived their entire lives strung between zero-G and the murderous press of heavy acceleration. Everywhere throughout the *Garveneer* was dotted with grab-handles and attachment points that paired with patches on our suits. And we had magnetic boots for walking around in, if you wanted to go fancy and retro. All this was our lot, facilitated by a ton of invisible, gene-level surgery and nano-implants which they started putting into us in the habitat tanks. Eyes, bone marrow, inner ear and the spatial centres of the brain. Weaning humanity out of the planetary conditions we'd evolved for, so we could gad about in free fall, or muscle through a protracted thruster burn without our bodies either collapsing or attenuating into nothing.

In the wake of Drone Fourteen's twelve seconds of fame, a special commendation came back from Chief Director

Advent over in Opportunities. We, the Special Projects crew, shared a tenuous moment of unity and cheer. You know, right before everyone went back to being frustrated at the planet's intractability again. Old Man Bartokh let his impenetrable solitude of rank slip enough to show some teeth in his grin. Skien and Rastomaier forgot they hated each other. Mikhail 'Big Mike' Jerennian, the data specialist, lost his punchy surliness enough to share a nod, and even bumped knuckles with me. Only Mai Ste Etienne was up and out of her seat straight away. Everyone shouted at her to at least sit down and enjoy the success.

'What?' Like Skien, Rastomaier and me, Ste Etienne was short and resource-starved. One of the many who'd grown up jostling elbows amongst the masses of the Concerns in the tanks, desperately taking aptitude tests until we fit some kind of mission-generated need. Bartokh's little barrel of a gut was a late addition, worn with pride to show how he'd bartered his natural gifts into the modest success of becoming a middle manager. Jerennian was just big, born that way. Some fluke of genetics that he'd fed with bare-knuckle cage fight winnings. Because when you grew up in the high-density orbital slums of the tanks, you made your own entertainment. Which made it extra weird he'd ended up specializing in fine data manipulation. But you can never tell, right? Anyway, even though Big Mike Jerennian had the mass of two of us put together, when Ste Etienne had her head set on something, he moved well out of her way. She was someone who didn't take no for an answer when she had a job to do.

Now she hung there, one foot in a strap and at a forty-five-degree angle to everyone else's consensual 'up'. She put her head on one side, rolling her eyes a bit. 'I thought you wanted a better drone. Not going to design itself.' I always had the impression Ste Etienne felt the rest of us weren't a necessary part of the mission process.

We persuaded her to sit down eventually, or at least Bartokh pulled rank. And, like all of us, any resource bonus Ste Etienne might have coming to her was strictly by his nod, just as he had to keep Advent happy to gain his own moiety. So after Bartokh had directed enough eyebrow-waggles and expectant expressions at her, she pulled her leg in and drifted to one of the attachment points. This meant I could take out the bottle I'd had printed at Bartokh's instructions. Something acceptably stimulating, but which wouldn't impair anyone's working judgement. I know most of the others abused their printer privileges to make worse and stronger stuff when they thought nobody was keeping track of usage statistics, but we were doing things on the official level, so it was going to be milkwater, as we called it. Hard to feel properly festive with milkwater, but we all of us raised a tube of it nonetheless.

As moments of success went, I took it. Nice to have a brief pause when I didn't have to watch everyone else for stress fractures. I hadn't asked to be the morale officer. I was just supposed to do what Bartokh considered beneath his dignity. Turned out that included actually making sure his team functioned. He wasn't what you'd call an approachable boss, and so I'd become the human equivalent of a passive-aggressive note left stuck to the wall. *Tell Skien that . . . If Rastomaier thinks . . . Ste Etienne has taken my . . .* Now seeing everyone actually united by some viable data left me realizing how crunched-up with stress I'd become.

They'd be at each other's throats again imminently, of course. Either from natural orneriness, or because some new demand would come down from Opportunities. *This too shall pass*: simultaneously the best and worst commentary on any moment of human experience.

Sure enough, there came a message ping with that special, slightly shrill pitch which told us it was from the top. Chief Director Advent was delighted we'd made such sterling progress. But Terwhin Umbar, from Resource Oversight, was appalled we'd basically just thrown away so much of the rare elements stock into Shroud's atmosphere, never to be seen again. Advent exhorted us to keep up the good work. *Ten more drones, a hundred, a thousand! Whatever it takes!* Umbar reined in our budget and expressed Serious Dismay about the waste. Where could we make economies? Did we really need all of six people to study an impenetrable alien moon larger than the Earth? Wasn't there a corner we could cut, so she could balance her precious books? Both these messages, each exerting its own inexorable gravity, arrived simultaneously in the same packet. It was time to put the bottle away and go back to work. Which, for me, meant trailing Bartokh to the small space he claimed as his sanctum.

Bartokh settled himself against the wall, the seat of his suit handshaking with the relevant point to fix him there for the duration. He had that expression of his, as though he'd just said something very funny which nobody around him was intellectually equipped to appreciate, but he was magnanimously forgiving us our ignorance. I had to remind myself that he was, in fact, extremely intelligent, and held his position as SP Director through genuine merit. He was *not*, in fact, just an egomaniacal pain in the ass. A gifted man, a man of vision and, if he was devoid of emotional intelligence, he was at least aware enough of the fact to make sure I was around to do it for him.

'Twelve seconds of usable video,' he mused. 'What's our trajectory?'

Meaning nothing to do with orbital physics, as I knew full well, but the far more difficult piece of navigation which would be threading the needle of competing demands Opportunities was saddling us with. What was going to happen next and how would it impact Umbar's budgets?

'Mai already has early plans for improved drones. She sent them over to me,' I reported. 'Mai' was Ste Etienne, because Bartokh was a First Names Person.

In order to act as the universal go-between of Special Projects I'd had to become everybody's understudy. I couldn't have created Ste Etienne's designs, but I could see what she was doing with them. I walked Bartokh through what she, and everyone, was doing. I was used enough to him calling these one-on-ones at a moment's notice that I always had the current checklist ready to reel off. Meaning this meeting could have been a memo. But as I said before, he liked his face time. An old man's understanding of why human contact was necessary. He didn't actually want much contact with the other humans, though, so I became the carrier wave for his atrophied humanity. Because Just Getting Along is the most important thing on the list, after air, power and food. Contrary to the old maxim, when you're all in the same tin can with limited elbow room, in space everyone can hear you scream.

I'd perfected a kind of brisk jollity for this task, like spoon-feeding an elderly uncle. Mai Ste Etienne and her drones.

Jerennian had declared more sections of Fourteen's recording as eminently salvageable. Skien and Rastomaier were working on new mission parameters so that Ste Etienne's improved drones would have something to do. And I, as well as acting as the glue for all the other elements of Special Projects, was working with Technical Oversight to make sure we had everything we needed to make all of this happen.

'Which means . . .' I ran Bartokh through a long list of choices – where we spent and where we saved, because these missions were always playing dice with entropy. And we were driven by a demand for results. Not just theoretical, oh-isn't-that-curious science, but a big fat report on Advent's desk entitled *Commercial Exploitation of Shroud* that could play into the wider development of this system. Another node in the greater project of human galactic expansion. Flags, trumpets, et cetera.

Big Mike Jerennian was the least congenial member of the crew. Meaning he still fitted within certain acceptable parameters, because nobody wants to be locked up in a spaceship with a psychopath. His particular neuroplot basically read 'monomaniac with little headspace for niceties'. Extracting a pleasantry from him was like drawing a useful breath out of Shroud's atmosphere.

You're not supposed to ever see your own neuroplot notes, but there's always a brisk trade in them via those who have access. I knew my own said I had an admirable flexibility, and should therefore be used to liaise between rigid people. Hence, as well as being Bartokh's emissary to the rest of the crew, and the entire crew's emissary to Opportunities, I was also the team's emissary to Jerennian.

I found him in his corner of the ship-belly workspace. He'd already flagged seventeen sections of snow-static that the algorithms reckoned contained retrievable footage. It was a delicate business. Both human user and artificial systems needed to make certain assumptions and predictions, and these affected what the cleaned-up images would resolve into. The human mind was very good at seeing faces where there were only clouds. And if you told the algorithms there were faces there, so were they. Jerennian was exacting, though. I almost felt the subjective human part of him got packed away somewhere in his head when he was working. I brought him a coffee-flavoured stimulant, and the grunt I received in return was parsable as a thank you.

After that, I retreated to the little crunch of space that was mine alone and set to work. Bartokh wanted me to create models of Shroud from the observable data. Not a good use of time, or computational resources, except that it would give Opportunities something to look at beyond the big resource deficit we'd built up making it this far. I constructed my phantom planet. I did it so well I almost *felt* it. The toxic, brutal conditions below us. The crush of the gravity. The icy bite of a temperature which never came closer than twenty degrees below the melting point of water, even on the . . . Well, not the equator. That wasn't how the thermal distribution of Shroud worked. It rotated precisely once per revolution around Prospector413b, the gas giant. Which meant one hemisphere was forever kept facing outwards into space, so that when the moon's orbit and its parent planet's were on the same plane, this hemisphere sporadically faced the distant star for a meagre injection of solar energy. Far more significantly for the moon's energy dynamics, the other face was constantly turned towards the turbulent roil of the gas giant itself, which put out considerable radiation and heat of its own. Light, even. A little almost-star that never quite could.

It wasn't as if anything would be sunbathing down there, though. No actual light ever reached the surface, and not just because the whole celestial business was way out in the waist of this planetary system. Shroud's atmosphere, the source of that informal name, was a turbulent yellow-brown fug, with levels of methane and other opaque impurities ensuring that below – that unknown surface Fourteen had fetched up on – was nothing but darkness. Those twelve seconds of life we'd glimpsed existed in an abyss as profound as the bottom of the deepest sea-trench.

By Earth standards, any life there should have been eking out an anoxic, chemosynthetic existence in a thermal vent. Nothing but microbial scum, with perhaps a few colony organisms building high-rise necropoli on one another's graves. Heavy, cold and dark. And crushed. While the atmosphere was thoroughly unbreathable and – if you tried, and before it killed you – would absolutely stink of urine and flatulence, there was also an awful lot of it. Which explained what had happened to some of the earlier drones, because the first couple hadn't been rated for that kind of pressure. So many of Shroud's parameters would gang up like thugs to kill you, but somehow it was the dark that upset me the most.

The dark, and the things living in it. Very unprofessional of me to think about a marvel of alien life like that, but there was something profoundly horrible about the idea of all that colourless, blind life fumbling about down there, groping through the methane-clogged murk.

With their radios broadcasting on all frequencies.

That remained an oddity neither Rastomaier nor anyone else could really explain. But it wasn't my problem. My models weren't trying to account for it. The whole Shrouded biome wasn't meant to feature in my work, save that it kept creeping eyelessly into my head. But all that upsetting ideation did spark a few thoughts. And, like I said, I was everyone's understudy, so had been forced to get up to speed on a very broad curriculum.

When I'd gone as far as I could with the data we had, I went to see how Ste Etienne was progressing. Her space was next to Jerennian's, and the pair of them generally just got on with things wordlessly for hours. Then – sometimes of their own notion, and sometimes because I came and bothered them about it – they'd stop, eat, drink, and discuss what they were doing in clipped abbreviations. Not actually asking me, or each other, for suggestions. But just because the act of putting their work into even those few words would shake loose new thoughts. I was their sounding board. Sitting there with them, chewing a nutrient bar, I wasn't sure whether they actually liked me, or each other for that matter. They had working relationships, for which liking was irrelevant.

That was according to the manual, of course. We had jobs to do, and a greater cause. There was a promotions and rewards structure that Opportunities dangled over us too. You did what you could to secure a hand on the ladder's next rung. You were a team player, until there was only one post to be filled and then you became pointedly better than everyone else. All the usual. Liking tended to get in the way.

I was very lonely.

Which is trivial, and not something that turned up in my reports. I didn't ever tell anyone I felt lonely, because that would have made me not a team player. I don't know if anyone else felt the same way, and if they did they wouldn't have told me, for the same reasons. Up there in orbit, we were all in our individual little tin cans.

In Ste Etienne's workspace projected screens overlapped in bewildering profusion. I saw a lot of drone schematics, already advanced beyond those latest she'd sent me. I could identify the sampling apparatus, annotated with a bewildering cascade of weight and thrust calculations, tackling just how to bring the extra weight back out of Shroud's gravity well. The composition of the atmosphere complicated any kind of regular rocket business. Everything would not burn at all, or burn far too much, depending on how much oxygen you tried introducing into the equation. We'd drawn Fourteen's lobotomized brain back using a hydrogen balloon warmed with non-combustive chemical heating, some serious fan technology, and compressed gas for that final Newtonian kick to push it high enough for our orbital scoops. All to recover a datastore about the size of my smallest toe.

Amongst the drone plans there was something else. I didn't understand it at first, because I hadn't checked the scale of the schematics.

When I dared enquire, Ste Etienne gave me a level stare.

'Your boss asked for it,' she said.

When people were pissed off, Bartokh became *my* boss rather than everyone's. I was the insulating layer between management and workforce.

Except this time – as he sometimes did – Bartokh had end-run around me and just sent instructions direct to Ste Etienne. Probably immediately on waking, before they fell out of his head.

'And this is . . .' I pointed at something, trying to pretend I knew exactly what was going on, and merely wanted some tiny detail clarified. Hoping whatever the detail was would clue me into the whole big picture that had been snuck past my back.

Ste Etienne wasn't fooled in the least. 'This is for the manned mission,' she said.

I should have nodded sagely but the enormity of it got away from me. 'The *what*?'

Ste Etienne shrugged. A roll of the eyes, a spread of the hands. *I'm just a poor engineer and all these dumbasses keep asking me to do the impossible.* Except there was a bit of a spark there. More enthusiasm than she usually let show. Devilment, basically. Devilment in an engineer is a terrible thing. The idea of sending a poor fragile human body down into that nighted hell actually appealed to her.

I bearded Bartokh about it. I explained at length all the risks this idea was running, with especial reference to all those ghastly, murderous numbers that had gone into informing my model. Bartokh said placatingly that, obviously, it was just a theoretical sideline, that we were still working with the drones and the instruments, and I shouldn't worry myself about it. Or about any of the other points I'd raised, where we were having to bite into my nice broad safety tolerances, because they made everything more resource- and time-intensive.

Bartokh then gave me that very clever expression of his

and, just as I was leaving, said, 'You know, Opportunities are really keen. Who knows what might be down there that we could use? If we can only take a good look. Drones and remotes, all very well, but limited, you know. Limited.' As though every other word we'd just shared had been entirely in my imagination.

1.3 LIGHT

The same broken cage of briars reared up on the screen. A point in time ninety-four minutes after the original twelve seconds, towards the end of Drone Fourteen's tenancy on the surface of Shroud. Although the drone itself was still down there, of course. After jettisoning its brain.

Its lamps lit very little. The colourless sheen of the arching, segmented stems, that looked more like plastic than wood or anything living. The faint flurries of the feeding fans or gills or whatever their function actually was. The limited range of the lamps the drone could mount barely cut through the sheer gloom, the curdled soup of what passed for air on Shroud. All was in shades of brown-grey, light and dark. Nothing had invested the energy into manufacturing pigments, because why put on an art show if nobody can see the pictures? Light and dark, and some yellowish tones, like old bone or diseased teeth or mustard gas. The brown of mud or excrement.

The view tilted slowly, and then righted itself as the drone's gyroscope registered the shift and corrected the extension of its legs. Another complication that had probably done for some of its predecessors, because if it wasn't pointed vertically at the end of its stay then there was no chance its tiny escape capsule could be propelled back into orbit properly.

* * *

'So, what we think,' I explained, at this point, 'based on the seismic data, is that we're actually over liquid here. The substrate is some kind of organic layer spread over . . . a sea. Or a lake maybe. The shifts we're observing are the motion of waves below.' Jerennian wasn't remotely interested in presenting his most recent findings to the team, so the task had fallen to me. 'Possibly something other than water, although the chem results suggest maybe water with high levels of impurity to keep it liquid. So maybe a lot of the earlier drones sank.' An awkward and expensive admission. We had been expecting ice or rock. 'We need to include flotation buoys on future drones.'

Ste Etienne grunted and made a note. I didn't say the other thing which had struck me. That there was no sign of water glistening in the lamps. The sightless sea I was hypothesizing was some unknown depth below the labyrinthine briar structure Fourteen had come to rest on. The thought of those depths, darkness layered on darkness, made me shiver once more.

In the next stretch of recording there was little sign of motile life. The big critter from the twelve seconds of wonder had gone on its sluggish, weirdly disconcerting way. But then, right towards the end of this two and a half salvaged minutes, something approached. Something even bigger, weaving its way through the briars. Many-limbed – it was unclear precisely how many because the whole thing was a confusing melange of moving parts. A giant monster that seemed hollow, like a wicker man. It advanced in a weird twisting motion, as though it was crippled, save that it moved with exaggerated precision. Articulated, multi-piece limbs unfolded, reaching and collapsing in a confusing, rippling sequence. The air glittered and danced, as though it was attended by a legion of fairies or fireflies. Then it shifted again and the lamps briefly caught the fine hairs gleaming all the colours the human eye could perceive – a prismatic lightshow that no denizen of Shroud would ever appreciate.

'Are they . . .' Hari Skien asked, screwing up their face, 'metal . . . ?' But there was no way for us to know.

The image grew more and more staticky, the trailing edge of the scrubbing efforts. Except the creature's approach and the inroads of digital noise were in lockstep. The front of it, a whorl of smaller limbs spiralling inwards from large to small, drew closer to the camera, and more snow veiled the image, but weirdly this was in time with the creature's blindly feeling advance. More limbs unfolded from within it, larger, their ends seeming jagged, raptorial, dangerous.

Static. Oblivion. Only seventeen more seconds, in fact, before the end of the recording, the launch of the drone's collected memories, and the end of the mission. This wickerwork monstrosity had been laid bare to human eyes, yet was too alien and complex for those eyes to quite understand. And it'd arrived only just in time to be seen at all.

'Tomorrow,' said Ste Etienne. She'd been fabricating pieces for days, her designs modular and customizable so that she could have three quarters of a drone ready to go, even as people were still making demands that would impact the remaining twenty-five per cent. 'The day after if you want floats on it.'

'We want floats on it,' Bartokh agreed thoughtfully. We were all kicking ourselves about that. A sea, at minus twenty to ninety centigrade, rather than the anticipated crust of ice. Which also meant that maybe all the work Skien had put into a drill rig for the drones hadn't been necessary either. More demerits the next time Umbar called to talk about resource allocation.

'How many do you want?' Ste Etienne threw in. At the surprise occasioned by this unexpected offer of bounty, she added, 'What? They don't throttle us until the next budget meeting. I know the dance. I've got a stock of parts ready for assembly. You want one, you want a dozen. I made sure we used what they gave us, so let's move it up a gear. If we're going to probe, let's probe the sonofabitch good.' Shrugging it all off as though it's nothing. Taking no credit.

'And samples?' Bartokh clarified.

'I mean, we're not going to go pull the legs off any of those sods,' Ste Etienne said. 'But we'll collect atmosphere, sure. Liquid if we reach it. Got the thrust equations sorted. Bring you back some bottles.'

Three quarters of the new class of drones didn't show back up for work after being dispatched. It was almost as if Shroud was a terrible, terrible place for anything even vaguely associated with humanity. In the end, though, we retrieved three usable data cores, along with three sets of atmospheric samples, and one that had slurped up a pipette of liquid.

Jerennian and his algorithms started sorting through the retrieved recordings, even as he continued to sieve Fourteen's records for any other viable moments. We all wanted to know what had happened in those last seventeen seconds, but Shroud had drawn its veil and the electromagnetic interference had peaked right as the drone launched its core. What happened when the thing from Shroud actually reached our drone? Love at first sight? Opportunistic and disappointed predation attempts? Did it hold up a sign with basic maths equations on it, in order to demonstrate its intellect?

Bartokh himself took over the atmosphere analysis, meaning he had long lists of jobs for me to do. We built a tank, the first of several as it turned out, engineered to replicate the enormous pressures of planetside Shroud. A big armoured fish tank we filled with murk. And in the murk, life. Microscopic life, certainly. No giant basketwork monster had crawled into the drones' tiny canisters. Every little capsule the drones brought back was clogged with it, though. The very air down there was fizzing with minuscule organic things. Everything else in the workbay was shunted over so that Bartokh could come get his hands dirty. He had his workstation, with a range of microscopes and analytical tools, and was trying his damnedest to work out what all that biomass was doing in the lightless, anoxic crush that was Shrouded air. Life needed energy. Big life needed big energy. And we'd seen that, down on Shroud, life could get very big. Bartokh's guess was that, rather than combusting stuff with free oxygen, they were sequestering oxygen in fats to combust with all that atmospheric hydrogen or something.

In a smaller tank, also under considerable pressure, was a fistful of Shroud sea, replicated from the sample we'd retrieved. Water with so much ammonia and salts dissolved in it that, whilst it probably did freeze across most of the planet, some of the time, it wouldn't do so reliably. Skien was already going into raptures about how the ocean dynamics would work, collaborating with me to throw up a series of weird and contradictory models we could show to Opportunities. And Opportunities was making noises that suggested there were commercial use-cases for this sort of stuff, if we could go down there to start exploiting it. Unique and alien biochemistry that could become unique and human-benefitting biochemical advances.

The water had its hitchhikers too, though fewer per cubic millimetre than the air. All the samples were a wealth of data waiting to be unlocked, if we could only figure out how the puzzle box fitted together. It won us another turn of the glass with Umbar's Technical Oversight committee. The asteroid-mined wealth of the outer system was slung our way so we could keep on investigating. Our wage-worth increased. We raised a glass of acceptable stimulant to the tenuous tightrope of our ongoing success.

Drones remained our main way of communicating with Shroud, a conversation aggravatingly one-sided most of the time. Our increased resource allocation gave Ste Etienne ideas, though. Specifically, she decided she was going to gift the moon with a space elevator, because we needed a reliable way of moving things in and out of that murderous gravity well. We had the materials technology to run a line of sufficient lightness and length down to Shroud's surface, if we could just find a secure attachment point down there. It would be an ideal way of countering the moon's lightless, EM-saturated conditions. What we didn't have was somewhere obvious to put it. Whilst the planet technically had poles in the sense of an axis of rotation, that sluggish once-a-year spin was tooth-grindingly slow by Earth standards. More of a problem was that much of this actual equator lay in the middle of what seemed to be high-energy weather systems, which shunted the atmosphere about in what looked like a constant hurricane cycle. In human terms, that rotational pole-and-equator set-up barely registered, and the key macro-features of Shroud in our minds became those inward and outward faces. We'd even started to talk about the two points facing towards and away from Prospector413b as 'poles', though Skien, the geologist, complained every time we did.

After a brief spit-balling exchange of messages on the topic, almost everyone else had given up the elevator as a bad idea. Apparently Ste Etienne had considered this an insult towards good Earth engineering, though, and became determined to set something up in defiance of basic orbital mechanics, survey data and our ever-threatened resource budget.

There was also the little issue that we still hadn't been able to confirm the existence of any solid ground for an anchor point most of the way along that rotational equator. The only definitive solid ground we'd discovered was (based solely on data from one very badly crashed drone) around the outwards-facing 'pole'. The region of the moon most defiantly pointing away from Prospector413b at all times – the very coldest point, where even the weather didn't want to go and there seemed minimal organic obstruction. There, we reckoned, was not only a prodigious depth of permanent ice, but solid rock below.

Based on this, what Ste Etienne jury-rigged up to create the elevator won us our third increase in resource allocation, and earned her so much back-slapping from the rest of us that she plainly wished she'd never bothered. Dropping a heavily shielded construction automaton into the gravity well was a terrifying gamble, which she only really warned us about after already committing the systems concerned past the point of no return. Given we had no way of communicating with the thing, there was a prolonged period when Ste Etienne was definitely nobody's favourite person, and we all thought she'd just screwed us out of a major chunk of our infrastructure for no reason at all. But then the compact orbital module turned up from below, with its nano-fine thread trailing behind it, gamely clawing its way out of the atmosphere and the gravity, bleating out for us to come and collect it.

That filament of a line was enough to move tiny construction robots down to the surface, carrying more lines, weaving the invisible into the slender, the slender into the robust, adding line after line, layer after layer, until we had something that could bear the stresses of orbit-to-ground transit. Between the foot-dragging rotation of the moon, and the outwards fling of it, swinging its way around the planet, plus a certain amount of brute-force artificial assistance, the line stayed taut. And this became our primary avenue into the hostile maw of Shroud. We could now shuttle counterweighted loads up and down into the lightless abyss of the place, take measurements and geological samples, and reliably retrieve as much weight as we wanted. Ste Etienne, who had a little more showmanship than Jerennian, actually demonstrated it for us, and for Opportunities. She obviously felt she'd got one over on the science faction and their high-minded malarkey.

The elevator also worked well enough for Opportunities to start planning out long-term polar resource acquisition. Later – and after our increased budgeting – one major problem with the situation became apparent, though. The outer pole of Shroud, being a bleak desert of ice and rock, was as devoid of life as the innermost reaches of Antarctica once had been, and as the majority of Earth now was. Despite all those atmospheric and geological factors keeping Shroud far warmer than an outer system moon should ever have been, Shroud's life apparently still found the extreme 'polar' – I can picture Skien sighing again at the inaccuracy – conditions inhospitable. Even the air of that outwards-facing region was depauperate, lacking the thick soup of suspended microbiota that more 'equatorial' regions boasted. So using the line to send drones down more reliably, and then balloon them out through the moon's skies, didn't particularly increase the rate of data retrieval. But then we didn't understand much about Shroud's weather patterns at that point. Ste Etienne took the news that she had not, in fact, solved every problem in one fell swoop with something other than delight. Instead, after playing with the elevator for a while, we left it to Opportunities to think up ways to use it, and turned back to dropping drones in the general vicinity of where Fourteen had gone down. Where we *knew* there was something interesting.

By then, Jerennian had found an easy-clean section of Drone Twenty-Two that he was ready to exhibit.

It was a miracle Twenty-Two gave us anything, because – Ste Etienne admitted in a rare moment of chagrin – it was defective. It didn't wait until it was on the surface to activate, but instead turned its lamps and cameras on during its descent. This serendipity confirmed an existing theory that the electromagnetic interference which attacked recording fidelity grew worse closer to ground level, because the first utterly uninformative minutes of the drone's fall were almost watchable. Showing, as they did, nothing but the roiling smoker's fug of Shroud's atmosphere. And then . . .

* * *

The drone was listing in its fall. Impossible to judge trajectory from the visuals but its instruments claimed it was being shunted sideways, its chute caught in winds of somewhere near two hundred kilometres an hour.

A brief flash. The image paused, then advanced frame by frame. A rockface, a mountainside, actual land, strung with vines or something fungal. The blurred suggestion of long straight lines leaned out into the tempest, conjuring unpleasant thoughts of some city-sized spiderweb. The drone whirled past all that, the view revolving vertiginously, and it was a miracle Twenty-Two didn't meet its end dashed to pieces against the cliffs.

Then everything calmed down for a second, perhaps in the wind-shadow of the mountain. There were maddening suggestions of shapes, just pale fuzz caught at the very edges of the lamps. Flying creatures, in those winds?

Later, Skien confirmed that the 'storm' which caught Twenty-Two was a fixture of the moon's temperature exchange cells, something regular we could model. The inner hemisphere of the moon, turned forever towards its planet, was warmed by its radiation. The outer hemisphere, facing eternally away from the planet, received none of it. The imbalance resulted in a constant dog-chasing-tail of a wind system, from warm to cold, and then the cooled atmosphere feeding back to the warm again. A predictable and constant tempest of circulating cells. Something life could evolve to exploit. Fliers, why not?

The winds whipped up Twenty-Two again, ragging the poor drone about like a rat in the teeth of a dog. If there had been a horizon then the view would have been a recipe for motion sickness, but all that could be seen was swirling fog caught in the lamps, glittering with particulate matter. Then came the mouths.

Pareidolia. Faces in clouds. For a moment the view through the drone's cameras showed the maws of dragons, lunging up on long necks to snap at poor Twenty-Two. Great round orifices, gaping in blind hunger. The image slowed again, algorithms doing their best to compensate for motion blur and focus. The suggestion of necks resolved into long threads, ruler-straight. The mouths – surely they were mouths – were big enough to swallow the drone whole, and their interior was a mess of complicated projections and processes, behind which was . . . the outside. In one brief frame, Twenty-Two's lamps and camera lined up perfectly to see all the way through, not a mouth but a tapering hollow tube. A windsock held in place in the raging airstream like a balloon on a string.

The kite-beasts of Shroud, living their best kite lives in the constant winds. One more mystery.

After such a protracted period of frustration and no data, the revelations came thick and fast. It was my turn to shine, to earn my wage-worth and give us something new to show to Opportunities. Just two days after the cleaned-up images from Twenty-Two had debuted, I gathered everyone who was still awake and replayed the footage from Fourteen. Those moments when the creature had been approaching, right at the end. Cross-referencing the timestamps of the video with what the drone had been picking up on the electromagnetic bands. We watched the creature's approach, again and again. Like a lopsided tardigrade made of wicker baskets, over five metres long. Each tentative approach – impossible for a human mind not to read curiosity into its odd motions – was paired with a sudden blaze of signal, rising far above even the noisy baseline.

'My theory,' I said. 'The interference we see comes from the creature. It's generating electromagnetic signals. All the life is. What we're hearing – sorry, what we're picking up from the planet – *is* the biosphere. Everything down there is generating a signal. Radio waves. Radar, basically. To navigate in the dark. To communicate, maybe. For all sorts of purposes, probably, because once you evolve something like that, I'm sure you can get all sorts of secondary uses out of it.'

Rastomaier eyed the stilled recording doubtfully. 'You think it's saying hello?'

He, Bartokh and Ste Etienne all looked at that inward spiral of projecting appendages, which might be a mouth or a face or neither. The last legible image before the static rode back in to reclaim its kingdom. To a human eye, there was only threat there. A predator about to tear up Fourteen's abandoned body, even as the drone jettisoned its memories into the higher atmosphere.

After presenting this, we received another injection of resources, because we were still in the phase of the project where discovering things was enough, and we didn't have to show just how any of those discoveries were useful. That phase was coming soon, of course, and we all knew it. I listened in on the world below, moving from frequency to frequency, hearing just its constant roar. Massed electromagnetic voices, all the living things of Shroud clamouring to be heard over one another. A dark world where shouting was the only way to live. And their actual audio would probably be almost the same. The first drones had been deaf, but Ste Etienne was working on the appropriate receptors. The atmosphere down there was dense enough to carry sound well.

Later, after I should have clocked off for my sleep shift, I ended up in Bartokh's office, where he used me as his sounding board. He had a lot to say about the atmospheric life, its metabolic pathways, and where their energy was coming from. Nothing down below was going to be evolving chlorophyll after all, yet the very air was certainly full of life.

'It's just . . . meatier than Earth,' Bartokh said. 'The atmosphere.'

I raised an eyebrow.

'I don't mean the suspended organisms but the *air*. It's mostly hydrogen-nitrogen, but there's a lot more complex stuff too: methane and ammonia, ethylene, acetylene. Big molecules. Organically useful molecules, just saturating the air . . .' Waving his hands about to help him think, as though he was shooing the offending atmosphere away. 'The air down there doesn't just *look* like soup, it's actually nutritious. Maybe. Except, you know what's happening in the tanks.' By now more drone samples meant we had several giant vivaria full of recreated Shroud atmosphere, in which Bartokh was attempting to grow his alien air-gardens. 'Just a huge ammonia build-up and then everything dies. Nothing does well in captivity. Mai's sending me more samples, isn't she?'

What came back with those next samples was the first Shrouded audio soundtrack. As I'd predicted, it was cacophony. The audible equivalent to the roar of radio traffic. Nobody could listen to it. And the only compensation was that the row on all those frequencies the human ear wasn't equipped to register was even worse. Jerennian set some algorithms to try and sieve it for meaning, but without much hope. Perhaps to a native it all fell into comprehensible channels and patterns, each species listening out for its own particular calls, but to a visitor from Earth it was an assault on the ears and nothing more.

I'd been pulling extra-long work shifts for a while, constantly in demand to smooth over relations between Skien and Rastomaier, or Bartokh and both of them, or Ste Etienne and everyone. Eventually my personal telltales informed me I was headed for a stress crash of my own, and I begged off for a double sleep cycle. The *Garveneer SP* shot me full of the appropriate drugs and I fell into blessed oblivion. I woke up later to an almighty row from the workspace because, apparently, without me, nobody could so much as zip up their suit without descending into an argument. Rastomaier and Skien had formed an unprecedented alliance to complain about Ste Etienne.

Once I was down in the workshop, the problem was obvious. It was barely recognizable as the same space I'd left. The real problem, honestly, was Bartokh, but nobody was raising complaints about him because he was their director, whereas Ste Etienne was just an engineer.

Bartokh's rack of vivaria were taking up a lot of room, along the curved side of the *Garveneer*'s belly, and originally they'd eaten into engineering's allocation of space. Now Ste Etienne had kicked off the sort of land-grab expansion that historically came with a cutting of diplomatic ties and tanks crossing the borders. Not pushing back into her previous territory, but shunting Skien and Rastomaier into the very hinterlands of the workbay, because her new project was *big*.

Three skeletal globes were hanging in the centre of the bay, held in place by lines that reminded me of the kite-beasts below. They were huge, easily as big as Fourteen's inquisitive monster. One of them already had an armature of four jointed legs, and two had modified gel couches installed within them.

I feel that I had a terrible and prophetic sense of dread on seeing them, but perhaps that's just hindsight talking.

'I keep telling her,' Rastomaier complained, 'nobody is going down there. Under any circumstances.'

Ste Etienne stuck her lip and chin out pugnaciously and kept on working.

'It's ridiculous,' Skien agreed. 'Half the bay. Half the entire space.' They threw up their hands. 'This is showboating, nothing more.'

I arrived in time to prevent actual physical violence on the part of the two scientists. Or maybe it was just that Ste Etienne was punchier than they were. The real problem was on his own sleep shift right then, which I reckoned he'd timed for precisely this reason.

I had to go wake him up, in fact, using a residual authority I think he'd forgotten he'd given me. Certainly, Bartokh didn't look pleased by the development when I told him, but he did look faintly guilty. A man who'd been hoping to awaken to a problem already solved or exhausted.

I cajoled him into showing his face down in the workbay, where he swayed with his foot anchored in a loop, positioned so that the new constructions were in his notional 'up'. Or only the scaffolding of them, right then – little more than blueprints given a third dimension.

Skien and Rastomaier rehearsed their complaints, while

Bartokh maintained his veneer of august detachment. As though he was, indeed, the impartial arbiter to all the disputes of lesser human beings. Once everyone had wasted far more time than was necessary, he admitted, 'Well yes, I did give the orders for these, of course.'

'You did what?' Rastomaier demanded.

'I told Mai to start work on them. As it happens.'

After he'd assured me it was all a theoretical sideline, now here it was, the main show in the big top. Manned vehicles for a human descent onto Shroud.

'The drones are too limited,' Bartokh admitted. 'We can't find out enough about what's below from piecemeal sampling. And without learning more, we can't program the drones for autonomous exploration. Too many unknowns. But we at least have enough data to design a Shroud-proof capsule. People are good at reacting to the unknown.'

'In that gravity?' Rastomaier complained.

'That's what the couch is for.'

'In that pressure?'

'Diving bells are old technology, Shemp,' Bartokh pointed out.

'Tell her to take them apart. This is ridiculous,' Skien said.

But Bartokh just stared up at the things, or perhaps at some vista of his own imagination. First man on Shroud. Ambassador to the wicker beasts. 'The work should carry on,' was all he said.

And so over the next month, as the drones came and went, and Jerennian restored recordings like an art historian with an old master, the manned vehicles took shape. Like great looming barrage balloons, they crowded out the workshop, giving everyone vertigo.

1.4 LIGHT

There were only maddening suggestions of movement in this snatch of footage. The substrate the drone sat on was particularly uneven and shifting. The view pitched and yawed as if from the deck of a ship. The beams of its lamps passed like searchlights over a landscape of chimneys reaching up past their range into the omni-present dark. Long stems – or perhaps the strings of those living kites – stretched from oblivion to oblivion through the fragile bubble of the drone's light. Beyond them, a wider-spaced lattice of basket weave existed, appearing more as smudgy finger-painting than definite presence. The one striking feature of this particular view was colour. The walls of the chimneys were streaked in reds and yellows, bright poisonous shades, but without pattern. No warning markings here, just some process that incorporated a chemical palette absent elsewhere.

'We know Shroud is volcanically active,' was my commentary, for the benefit of our distinguished guests. 'We think Thirty-One came down somewhere near a rift, a vent. We think this is a . . . community that's feeding off the relative heat –' meaning a few degrees higher and still well below freezing – 'as well as the unique chemical resources to be found here.' I could hear my voice wavering with nerves.

My interlocutor, the audience this was all aimed at, said,

'This is underwater?' Frowning at the blurry, indistinct images. It could have been underwater. Maybe I should just have affirmed that and moved on, but pedantry got the better of me.

'This is in the open atmosphere. In fact, it's directly over water. Or the water mix that makes up the seas. We're seeing structures built by Shroud life over or out of the sea, above the vents.'

'What's it for?' he asked, and I had to say, 'Well, we don't know exactly what it's doing.' I was losing him, so I played our hole card, the big thing.

There was movement: something. The algorithms reckoned it was about a metre long, so small to mid-sized for Shroud. It was creeping around at the edge of the light, picking its way between the fuzzy, indistinct strands. There was a suggestion of very long stilt legs that worked weirdly. Everything a little offset, as though there was a visual glitch. That strange pause between each operation, and then the motion itself very swift and precise. A stop-motion world. Then it was gone into the darkness, leaving nobody the wiser. The images sped up, hunting the recording for the next treat, to when this something came back. Just as indistinct, going about its business without any thought for who might be watching. Not refusing to step into the limelight, just not knowing there was such a thing as light, lime or otherwise.

'So, here's this thing, whatever it is,' I explained, and my audience, the important man, nodded with a slight air of *Is this going to go on for much longer?* I hurried on with, 'And here's our plot of the signal interference. We've begun to isolate distinct patterns, you see. Or at least sometimes.'

Bartokh would have preferred me to oversell what we'd actually achieved, but I couldn't make myself do it. 'This is its voice.' Identifying a jagged plot, I highlighted it in colours that were hopefully cheerful and not poisonous. 'You can see how the signal strengthens as the creature, this alien, comes closer.'

'What's it saying?' the man asked. He actually sounded interested for once, but of course it was the worst question, because . . .

'Well, not *saying*. We think this is its sensory pulse, that it's using to "see",' with little finger quotes, 'its surroundings. It's generating actual radio waves biologically and bouncing them off everything around it. Now you see here . . .'

A section from a different recording flickered up. Cheating, really. A moment in an otherwise utterly uninformative minute of film when something clattered past the camera's eye. There was no sound, but a human mind inserted the clatter because of the visual cues – the plates and interlocking struts of the thing's outer substance as they rattled and jolted. Faster than most of the deliberate motion previously observed, and just a blurred glimpse of it. Five assemblages of rods and joints caught mid-reach, as though someone had decided to build a dinosaur-sized pillbug out of twigs and string. Then the image froze. An overlay highlighted a pattern that scattered across its sides. Not colour, exactly, but composition. A dance of metal inlay streaked in jagged weaving daubs.

'What we *think*,' I said, desperately trying to infuse the word with far more certainty than it was designed to bear, 'is that this is a sort of baffle pattern. It uses shape and material to give misleading results about what and where the organism is, when you hit it with a radar pulse. Camouflage.'

'It isn't working,' the important man observed, sounding amused.

'Not with *us*, but nothing on Shroud has eyes,' I pointed out, surprising a thoughtful look from him, bringing home the true alienness.

We were, in a way, victims of our own success. Opportunities decided to join us, relocating from the outer system, in fact, because Shroud was proving such a unique source of data. The rest of the *Garveneer*'s work could go on almost remotely, with a handful of mining teams overseeing an army of auto-miners to generate resources, which could then be distributed to everyone else with a lordly hand.

So the main hull of the *Garveneer* now spooned around our Special Projects module, supplanting our solar array with its own, and causing all the usual problems as nothing quite linked back up properly. Everyone received a dozen different demands from various departments, requesting data, assistance or advice. Abruptly our leisurely research work on Shroud was stepped up by a need for commercial exploitation.

Sharles Advent, Chief Director of Opportunities, had come over in person to see my presentation and tell us what a good job we had all done. He was an unremarkable-looking man, honestly, save that he was stockier than anyone except Jerennian. A man who'd been well fed from childhood, rather than scrabbling for scraps in the habitation tanks. He listened to everyone's little bit about what we'd discovered and what we'd accomplished. Then he leant back and exchanged a few words with Umbar, of Technical Oversight fame, who'd accompanied him. She was a pleasant, plump, open-faced woman and we were all terrified of her. Advent made the big decisions and Umbar decided how they could be afforded. If that meant your continued waking existence became commercially unsustainable then it was the hibernation shelf for you. Whenever she spoke back to Advent in her soft whisper of a voice, every word rang with the refrain of, *Yes, but where does all of this get us? What's the bottom line?*

Bartokh had the final word. He waxed expansive on the topic of Shroud being an unprecedented opportunity, a gold-mine of never-before-seen organic interactions, surely a treasure chest of biological processes we could use for our own purposes. Everything down there was unique. Everything was a lesson which might revolutionize some branch of human knowledge, large or small. Umbar's look back to him said plainly, *But you don't really* know *that, do you?*

'Well,' Advent said at last, and with at least a veneer of thoughtfulness, 'we know there's complex chemistry down there, organic and non. That makes establishing a harvesting foothold worthwhile. The elevator's a success, so we can at least reach the surface and communicate with it via the shielded fibre-optic line you've rigged up. As for the rest of the planet.' His face twisted. 'Doesn't make it easy, does it?'

Bartokh was forced to admit that Shroud was indeed an uncooperative partner in the venture, but one we had every expectation of bringing into line. This was our cue to shift from the social space, where we'd been sitting and playing our home movies, and go down to the workbay. Where the hanging fruit of the manned vehicles could be showcased.

Advent's face twisted very slightly when he saw them. Exactly as mine had, I suspect. Beneath that mask of rather vague bonhomie there was a mind which had absolutely understood how hostile conditions were below.

'You'll be looking for crew, I imagine,' he said.

'Eventually, Chief Director,' Bartokh agreed. 'I'm sure we have some people in store with appropriate aptitudes.'

We exchanged glances, we Special Projects people. I suspected that 'appropriate aptitudes' included extreme expendability. The dregs of the *Garveneer*'s slumbering human cargo, not woken yet because nobody had any use for them. People desperate enough to earn some basic wage-worth that they'd volunteer for something as mad as going down to Shroud. A monstrous prospect which, I can only stress, no sane person would ever consider.

Chief Director Advent looked at the now mostly complete vehicles above him, the layered thickness of their hulls, the arthropod clutch of their quadruped limbs. His eyes shone with frontier spirit, on the very reasonable basis that it wouldn't be him braving that particular frontier.

'You've had more than your share of resources we've extracted from this system, Director Bartokh,' he noted mildly. 'And we don't begrudge you it. This is a remarkable opportunity and your team has risen to the challenge adequately. However, we need to step up the pace. Keep on with the drones, certainly, but accelerate. We don't have a lifetime. All this –' a hand waved at the specimen tanks, the workbay, all of us – 'needs to start yielding actionable results before the next review. I have more than one mining team director asking if perhaps your moon here shouldn't fall under their purview, you know. I really don't want to get involved in that kind of spat. So make sure you have something to show for yourself, won't you?'

* * *

I should mention that Advent and Umbar weren't alone the first few times they dropped in to see which targets we weren't hitting. At the start there was a fussy, nervous-looking third person called FenJuan, who everyone thought was going to be given Bartokh's job. Bartokh obviously thought so too, because he monopolized far too much of my time asking me to look into their records and find out what their competencies and neuroplot were. FenJuan was, for a brief time, over-enthusiastic and a bit of a nuisance, because they were a xenospecies specialist and had a mad amount of keen about the possibility of alien life, out here where nobody had expected it. For a moment it looked like the whole circus was about to go seriously left-field with a new ringmaster, and who knew what tricks we'd each have to learn?

All of this became moot, though, after FenJuan suddenly wasn't around any more, and it turned out they'd been shoved back on the shelf. A Big Theory type who would probably have been a really useful addition to the team, but they'd gone too wide-eyed when they saw our drone footage, and Jerennian said later he'd heard them arguing with Umbar about things. Important things. Mission objective-level things. It wasn't some political masterstroke by Bartokh that had FenJuan, whoever they'd been, consigned to the hibernation couches again. It was a demonstration that nobody was invulnerable, and the nail which stood up would be hit. Possibly they'd thawed the poor bastard out specifically to make this point to all of us, and to Bartokh most of all.

The next month was trying, as Opportunities and a dozen subordinate departments moved in and pushed for elbow room. I became the vanguard of a whole fierce war of coldly polite remote messaging, as Special Projects defended its workbay from the encroachments of various ambitious factions. Only our existing successes to date allowed us to hold our borders. I spent half my work shifts closeted with Bartokh, as though he was a general moving pieces across a map. It helped that Ste Etienne had a kind of stare she deployed on anyone who tried to muscle in on her space, which tacitly suggested she was imagining killing the interloper with a cutting torch. It also helped that Big Mike Jerennian was genuinely very big, an evolutionary throwback in a population of rake-thin habitat tank kids. I felt vaguely offended that this did help. It shouldn't have, but it did. Also, the influx of newcomers brought several others who'd done the cage-fighting matches, back in the entertainment-starved days. Jerennian didn't get into actual fistfights over work shifts and resource allocation, but he did break a few heads for illicit entertainment, winner takes the surplus. All of which was strictly against *Garveneer* ship rules, but I swear I saw Advent himself in the audience one night. I didn't think I'd ever stand in a tightly packed ring of yelling crew, shouting for my work colleague as he knocked the pips off an assistant line manager with pretensions of kung fu, but I damn well did. It felt weirdly cathartic too. Better than another round of snotty messages, anyway.

We were under the hammer every moment. The orbital construction team was building the great skeletal frame of a full-on commercial exploitation hub. From this the Concern would be able to coordinate the system-wide effort to harvest, build and make Prospector413 one more hub in humanity's outwards expansion. A place where the next wave of ships from Earth could repair and resupply, before heading outwards yet again. And from here, the theory went, we'd also deploy a thousand machines to Shroud planetside. Miners, refiners, harvesters, processors. Or we would as soon as we could get a better handle on what was down there. Our deadline for actually presenting our final report, *On the complete and unobstructed asset stripping of Prospector413b Moon I*, was approaching at terminal velocity and we had to jump a lot of hurdles faster than anyone liked. Every iota of data became gold dust, and each time the slightly bruised Jerennian cleaned up another section of recording, everyone who was on work shift gathered round it. Which meant four or five of the six of us by now, because Bartokh had rejigged the schedule and cut everyone's sleep allocation. Drugs all round to keep us sharp. And, when you were on those sorts of drugs, the stuff down on Shroud looked *really* horrible.

There were several of them. The fields of the drone's cameras didn't overlap and the alien animals all looked the same. When they moved together it was hard, in the stark lamps, to see where the wickerwork structure of one ended and another began. 'Several', therefore. But just one at first. It had arrived and then gone away again. And then, in a later section of clear footage, it returned with friends.

Rastomaier had been in ecstasies when this happened. Even with Opportunities breathing down our neck for usable exploitation strategies. 'Look, cooperation, social interrelation. Complex behaviour,' he crowed.

'You don't know that,' said Skien, the eternal counterpoint. 'Maybe the others just followed this one in case there was food.'

'Why to the drone, though?' was my question.

'Maybe they eat drones,' said Skien uncharitably. 'I mean, we reckon this one ended up near where Fourteen was. We've dropped a load in, aiming for the same area. They could have developed a taste.'

The creatures were never quite still on the footage, though their movements were deliberate and patient by Earth measures. The flexing of their bodies plucked rainbows from the drone's lamps as their metallic hairs caught the light. We thought they were the same manner of beast that had come for Fourteen right at the end of its recording. Now they were here to pay homage to Drone Forty-Eight, the latest visitor from the stars. They jostled and humped about on their puppet legs. They didn't stare – they had nothing to stare with – but a battery of alien senses washed over the area with especial reference to this metallic intruder.

I'd isolated several transmission bands they seemed to be using. Some I pegged as definitely sensory. Forty-Eight's upgraded instrument suite reported being bombarded with a variety of electromagnetic and sonic pulses, many of which had to have been narrow-targeted rather than just a wide pulse, based on how energetic they were. The beasties were locating Forty-Eight by fielding the bounce-back from its hard, artificial exterior. We couldn't know what level of analysis was going on in their alien frames, but this was something they'd never come across, and it was impossible not to read curiosity in their constant barrage of pings. They were, I believed, fascinated. If they thought at all, what must they think?

There was also a wash of other signals, waxing and waning with the creatures' presence. Not specifically aimed at the drone, but peripherally detected by its instruments. Rastomaier and I dared to suggest it was evidence of intraspecies communication. Sophisticated communication. Like the dance of bees or the songs of birds.

Nobody else believed a word of it.

The things approached the camera's eye closer. Long flexible whips tapped and caressed at the drone's sides, seeking tactile feedback. The attempt kept one creature still enough for the image to resolve how it was put together. A series of overlapping basketwork segments, each sporting a jutting, multi-piece limb, every section of which seemed to have projections feeding deeper into the beast. Levers. Individual struts surely too thin to contain any analogue to muscle, or perhaps any living tissue at all.

'You're saying it's a machine?' Skien demanded incredulously. 'You're saying some alien worm-monster down there is building giant walking mech-suits or something?'

'No,' Rastomaier said patiently. 'But if you have a better explanation the floor's yours.'

'How many legs?' asked Bartokh, who until now had been content to let his cohort bicker.

'Yes, well,' I admitted, because we'd cut mention of this from our reports so far, to avoid ridicule from other departments. 'In this species we think seven.'

Seven legs, one per segment, on alternate sides, like the leaves on a plant stem, giving the creatures their bizarre rocking gait. Yet their environment was a giant briar patch and they picked their way into and out of the lamplight with measured assurance. They could also huddle together without ever quite touching, and reached out towards the drone again and again. Then one, without an obvious prompt, started approaching even closer. Its spiral of forward appendages flexed and unfurled like a flower made of cutlery. Static washed back in. Four seconds later, Drone Forty-Eight spat its memories upwards for later recovery. Nobody wanted to think about what happened to its body left down below in the dark.

'The thing is,' Advent told us, 'we're on a tighter time-budget than you realize.'

We were in Bartokh's space. Advent, Umbar, Bartokh and me just about squeezing in. Advent was the smooth operator, the gladhander, always eminently chill so that he could tell you you'd need to do twice as much in half the time and somehow make you look forward to it. Management skills. Umbar wasn't. She was his attack-dog, basically. I wondered if, secretly, she was as resentful of her role as *I* was when I had to play human-contact intermediary for Bartokh.

'We are making considerable strides,' Bartokh said, matching him genial for genial, mild for mild. 'Surface conditions are yielding to our instruments. The chemistry is mostly an open book. The actual engineering challenges of establishing a long-term presence, manned and automated, are all falling into place.'

'We're delighted with your work, obviously,' Advent said, with one of those smiles that went no further than the outermost dead skin layer of his face. What he meant was *You promised too much and we're here to collect*. 'But we need to scale up resource intake on all fronts so that we're ready for the next wave.'

Bartokh and I exchanged looks. My superior adopted an expression of polite enquiry. 'Hmm?'

'Director Bartokh, this system is ideally suited to become

a jumping-off point for future expeditions. We're not just stripping and leaving, or setting up a remote refuelling post. We're building for the long term. A shipyard, fuel, maintenance, supply. This system's going to be not just a roadside stop, but a new expansion hub. A jewel in the Concern's crown. That means we have a great deal of work to do. Umbar wants to wake the ground mission crew. Except she can't really train them without you giving me just a *little* more hard data on conditions down there, can she?'

'I see how that presents an obstacle,' Bartokh agreed. 'If you could perhaps green-light my most recent resource requests . . . ?'

'You do appreciate we have other departments?' Umbar cut in, on cue. 'You're already monopolizing a disproportionate amount of everything we're bringing in. Get it done with what you have, Director Bartokh.'

When they'd gone, Bartokh rubbed at his face. 'You've seen the red folder,' he said.

I had. It was where we put all the concerns Ste Etienne and the others had raised, about how close we were cutting the safety margins. Right now, the whole operation felt like a clown car bursting at the seams, ready to spill all of those brightly coloured jolly entertainers out into hard vacuum.

'Delete it,' Bartokh said. It wasn't even the first time.

But it wasn't that which did for us. In the end, although we cut every corner there was, scrapped every safety measure in our haste to make Advent happy, it didn't even matter.

1.5 LIGHT

The last twenty recorded seconds of Drone Forty-Eight, before it launched its tiny memory, its camera was full of the shivering, shimmering, eye-confounding alien beasts. Each one of them a puzzle, plates and pieces sliding over one another, pierced by holes that gave onto depths where more pieces slid and moved. It was impossible to know if anything that could be seen was actually the living thing at all, or just more accreted pieces, like the case of a caddis fly larva.

Ste Etienne had been playing the sequence continuously; obsessing over it while the next round of pieces for the manned vehicles was printed out and fitted. Whilst a final layer of flexible glass-foam was pumped into the gap between shells, which would give the whole pod a flexibility against impact whilst maintaining its pressure resistance. Because going down onto Shroud's surface would be like plunging into the deepest ocean, up to and including the possibility of attack by monsters.

A sinuous shifting of plates as one of the hosts approached the drone. A build-up of interference from its inquisitive signalling, exploring the dark world with electromagnetic feelers that returned their information at the speed of light. The images briefly seethed with snow and then settled as the scrubbing routines compensated.

The background of the recording was just sections of monster shifting past one another, but in the foreground was the best view yet of a single creature as it reached for the drone. That blind spiral face opening out like a fractal, always with more and smaller arms unfolding from its heart. To a human eye there should have been some commonality there. They were not so infinitely alien, surely. And yet the blindness of them, the weird asymmetry of their bodies, the bizarre intricacies of their construction, like mechanisms, like toys, all spoke of a queasy wrongness.

Jerennian's algorithms had allowed for the isolation of one beast amidst the rest. Still unreliable and difficult to decipher, because they were large enough to extend beyond the corners of the camera's eye. And because they were always in motion, with their surfaces like the outsides of pollen grains magnified thousands of times, all those whorls and holes, teeth and crenulations. Armour; casing; components; clockwork. There was an old, old phrase about the philosophy of evolution, the argument over how things arose from a primordial nothing. If ever there was a world made by a blind watchmaker, it was Shroud.

Every time I looked over at her, I saw Ste Etienne still squinting, enhancing and magnifying as she tried to work out what was bothering her.

Looming close, so the camera fought for focus. The arms unfolding, large to small to smaller, and never quite seeming to reach smallest. Reaching hungrily for the drone, a meal of metal for creatures that could obviously metabolize it. Greedy, mindless, hungry. And yet . . .

* * *

Then the next series of components for the pods were ready, and she had to break off from her personal research. I tried to snare her with an interrogative look, but the engineer was keeping her secrets. Never the most communicative of the crew, right now she was closed as a sphinx.

We were all at it, though. Yes, the pressure, the unreasonable demands, the straitened resources, but at the same time, Sharles Advent had done his job well. Turned up, smiled, congratulated, shaken hands. Made us think of the bonus, the enhanced wage-worth, the potential promotion if we pulled it all off. Now it'd been revealed that Prospector413 would be more than just an exploitation centre, we were dreaming about the new posts that'd be available. People were coming to *live* here, and that meant a whole load of positions with a better balance of reward and responsibility. Everyone had a side-hustle designed to show how they were so much more capable than their record to date might indicate. Except me, honestly, because holding the entirety of Special Projects together under Bartokh's laissez-faire management style took up every waking second of my extended work shifts.

Under Ste Etienne's direction, Skien and I helped fit the new pieces, then injected more foam layers. We set up the testing algorithms, running through system after system to ensure everything was shaking hands with its neighbours properly. Ste Etienne hovered by the readouts, watching not the actual data but the intelligent summary of it that her aggregation systems presented. Skien then returned to their own data-sieving, but I remained at the engineer's shoulder, because *this* had become the heart of our venture. Beginning the exploration in earnest. Putting a human on Shroud. One whose only link to orbit would be the shielded fibre-optics of the elevator line, and who would have to maintain an unbroken link to that if they wanted to receive any new instructions, or relay any data. Or call for help. Otherwise they'd have to strike off into the unknowable wilderness of Shroud, entirely self-sufficient.

Eventually Ste Etienne grunted, which meant she was satisfied. Then, before I could go report to Bartokh, she said, 'I need your opinion.'

This was unprecedented. I was, after all, the understudy. Nobody wanted *my* thoughts on anything, they just wanted to make sure *I* understood *theirs*. And Ste Etienne was far from the most patient of teachers. She thought I was a bit of a fool, Bartokh's creature, or that was how I'd always sensed it.

But there was something in Ste Etienne's manner that registered with me. A shiftiness, almost. The fact she valued my opinion less than anyone else's also meant she was limiting potential embarrassment if she turned out to be wrong. There was something up and it was sufficiently odd to be left out of her reports, yet sufficiently intriguing she couldn't not mention it.

'Look at this.' She had the recording up again, advancing through the frames until the image was nothing but that whorl of reaching limbs – it still sent a little shudder down my spine. She fiddled with the focus, letting the algorithms do their best to bring different elements of the creature into sharp relief. 'There. Look.' Awkward, suddenly, at having to put it into words. 'It has to be an artefact of the image enhancement, surely.' Her look at me was uncharacteristically naked and vulnerable. 'Am I going nuts here, or what?'

'What am I looking at?' I frowned at the image.

Ste Etienne highlighted a handful of points, the termini of some of the mid-sized limbs in that whorl. She then stared impatiently as I continued not to see it.

'Look. Wait.' The engineer kicked off across the room – she was by far the best of us at doing those zero-G acrobatics. Everyone else preferred to crab-spider around the walls, loop to loop, hand over hand. I kept on staring blankly at the image until Ste Etienne arrived back, brandishing something. 'Here, standard attachment deconstructor tool. Fancy name for a spanner, basically.'

I looked at the metal implement, a piece from one of the construction waldos that had been doing the actual lifting and building in the workshop. I knew what it was, of course. I wasn't so very clueless. I'd been trained in its use, even. Before Ste Etienne's doubting gaze I opened my mouth to make some general declaration of bafflement.

Ste Etienne held it up in front of the screen and finally I saw.

One reaching limb of the creature terminated in the same shape. Not entirely identical, but the inner crook of its surface had the same topography as the business end of the metal tool Ste Etienne was brandishing. A shape designed to fit exactly the fastenings of the drone it was reaching for.

For a moment I was completely adrift in a universe I had lost hold of, as though I'd kicked off from the floor and was floating away across the workbay. *Impossible.* Surely it was impossible. Of course it was.

So there was only one explanation. Occam's razor.

'You're right,' I said. 'It's an artefact. It's got to be.' The algorithms, in trying to enhance the image, had made a weird leap of assumption and decided the static-sheathed tips of the thing's limbs fitted something from entirely the wrong image set.

'Right. Of course,' Ste Etienne agreed flatly. 'Obviously.' Trying to sound relieved, but her usually steady voice was shaking a little. Her look was almost pathetically grateful, swiftly papered over with her preferred uncaring shell. She shrugged off any further attempt of mine to engage with her, retreating into the nearest of the globe-shaped vehicles to run through more system tests. I continued to stare at the images and wondered what kind of AI pattern-matching blunder had caused this, and how that impacted on the reliability of everything else we'd seen.

I didn't have much time to ponder it, before The Accident. Which wasn't even our fault.

Amazing, honestly, that it wasn't. We'd been burning the candle at both ends ever since Opportunities moved in and put us on the clock – going mad to get the manned craft ready. We'd cut all manner of corners on gathering data from the planet below, meaning that instead of actual conclusions, we were presenting a range of fanciful suppositions and hoping they were ballpark-close to the truth. There's the same intractable triangle that sticks itself like a shiv into any project, really. You need things done quickly, and cheaply, and well, but if you want them quicker, or cheaper, or better then one of the other things has to give. Advent sent around cheery little messages every day about how pleased he'd be if the pace picked up a bit, while Umbar sent dour little notes informing us that we didn't really *need* all those expensive tests and redundancies. Why, spending resources on redundancy, failsafes and backup systems was as good as admitting you weren't good enough.

We weren't good enough, and neither was most of what we built or reported. Through that last hectic month of frantic work, five on-shift and one sleeping at any given time, we never even considered that all the other departments were working under the same ideology.

I was in the workbay when it happened. Most of us were. It was Jerennian's turn to lie in, which was what made it weird when he called in to us.

'Why am I hearing alarms? What's going on?' Just his voice, because nobody could spare the attention to look at his tousled, sleep-deprived face.

'There aren't any alarms,' Skien told him dismissively. 'A nightmare. Take drugs. Go back to sleep.' By then we needed drugs to sleep, and then drugs to keep us awake. Umbar was complaining about that element of our resource usage as well.

'There *is* an alarm,' Jerennian insisted over our comms link. And we could actually hear, as the backdrop to his words, an alarm. Quite a shrill one. Then he made a kind of strangled sound, and what came next from him was garbled as the alarms redoubled and washed over it. I only caught the word, 'Outside.'

Bartokh rushed in then. He'd been in his own office, his little workspace. He was wide-eyed, clutching for the straps as the door irised shut behind him. 'What is it?' he demanded.

We looked at him dumbly.

'The alarms,' Bartokh said. 'What?' And he visibly registered there weren't any in here. Not in the *Garveneer*'s Special Projects module workbay. Because, I can only assume, someone on the main ship crew hadn't connected us up to their system properly when they rejoined us. Funny what you only learn when it's too late, but then that's built into the concept of problems with emergency alarms.

'Uh . . .' Just a sound, from Ste Etienne, but she'd caught the *outside* from Jerennian too, and called up some screens. Her eyes went very wide. 'Fu—' she said.

I couldn't understand what I was looking at first off, but Ste Etienne had an overlay on the image a moment later, presumably the one Opportunities was using to try and limit the damage. The damage that was about to hit.

I could see the *Garveneer*, meaning the main vessel, and the Special Projects module we were in, which it was clasped around. There was also the extending scaffolding of the orbital hub that Construction was throwing up as quickly as possible. It was mostly just skeleton still – long, curving struts and ribs and the occasional workspace module bolted on like a wart. That was resource-intensive work, of course. They needed a lot of raw material, and it was coming in from the asteroids, as well as some of Prospector413b's other moons. The mining was established practice, the first thing anyone started on entering a system. They'd been retrieving useful mass and elements since before we were woken up.

The way it worked was like this. Each consignment of mined material was loaded into a mass catapult and accelerated towards us, using the sort of precise maths you can do in hard vacuum, even over the enormous distances involved. Then there were interceptor automata which would latch onto the shipment and decelerate it, until it could match our orbit. Simple.

The interceptors came in for a lot of rough treatment as you can imagine. They were on and offline all the time, being fixed up. Which was fine, because they signalled their readiness to the catapults before they launched. I mean, we weren't fools. Also, they made sure the catapults weren't ever launching anything directly in line with the *Garveneer*, just in case. Even with all the safeties we'd end-run around, that was still plain common sense.

Of course, the enormous skeletal spread of the new orbital hub was a lot bigger than the Special Projects module had been. And maybe one of the catapults was set up slightly out of place, without its instruments registering the fact. One of the interceptors was also reporting a fault, after previously signalling its readiness, but there was no point telling the catapult that because the consignment was already on its way, hurtling towards us with inexorable, Newtonian force.

There was about to be an equal and opposite reaction.

'It can't be coming close enough to be a problem,' Skien said.

'It's going to hit the scaff,' Ste Etienne pronounced, almost awestruck.

I looked at the images. The impact point was at the very furthest extent of the orbital scaffolding. Barely a brush, really. I could see the distances changing as the *Garveneer*'s engines came online and tried to nudge the whole show very slightly, to tuck us out of the way. Around us, the hull thrummed with gravelly vibration.

'I'm going to back up the data just in case,' Rastomaier said, because that, apparently, was his priority. He kicked over to his terminal and began inputting commands.

Bartokh licked dry lips. 'Um . . .'

'Into the pods,' Ste Etienne decided. In fact, she pronounced it from her new position clinging to the side of one, because she practised what she preached.

'What?' I demanded. 'No.'

'They're engineered for a crapton of stress,' she told me. 'They are the safest thing in this place. Get in.'

I looked around for Bartokh, because I wanted him to tell her not to be so stupid. Nobody needed to get into the pods. That was an entirely different sort of terrible event and could we just deal with the problem we had? But Bartokh had slipped his feet out of the deck loops and was already up on the second pod, clambering towards the top hatch. Skien bent their knees and launched, flailing, towards the third. They bounced off it, landing spread-eagled against the upper wall, then pushed themself determinedly back.

'No,' I repeated. 'Come on, now. No.' Bartokh and Ste Etienne were already inside by then. Skien had got themself at the hatch of his pod and was calling for Rastomaier to hurry up. Ste Etienne's head popped up again and she was shouting at me. She was actually screaming at me, saying something about velocity and time.

I had not understood. I'd looked at that serene image, the vast bulk of the minerals shipment cruising through the immensity of space, and hadn't actually checked on any of the important details, such as how long we had before impact.

Over the link, Jerennian was saying something, tense, alarm-occluded. I didn't catch the words.

So I leapt. Went wide, but Ste Etienne lunged to full extension and snagged my ankle, yanking me back. I crashed into her in a snarl of limbs.

She had a screen up inside the pod too. I saw the shipment impact. So delicate a touch, not enough to visibly deflect the thousand-ton man-made asteroid as it went on its course to oblivion. And nothing happened.

Nothing happened subjectively, for precisely the brief period of time it took for the vibration of impact to travel along the orbital frame and reach the *Garveneer*.

Some godlike hand slapped the entire ship from one side. Ste Etienne and I were thrown into one curved side of the pod. It was cramped in there, and this stopped either of us building up a bone-breaking momentum, but we'd have a whole crop of bruises from it. Then we were thrown the other way as the pods impacted with the side of the work-space. I realized the high singing sounds I could hear were the anchoring tethers snapping under the stress.

The noises went on, the shudder and ripple of the shock-wave convulsing through the ship and then on to the furthest extent of the scaffolding. And then came back, reduced, the kinetic energy spending itself in diminishing violence.

Ste Etienne and I looked at each other. Her nose was bloody, enhanced clotting already sealing the flow. She pushed me off her and scrabbled for the hatch.

We heard Skien's voice when we opened it. They were shouting for Rastomaier again. Popping my head out revealed a scene of utter devastation. Everything that wasn't bolted to the walls had been overturned, and was now floating, spinning in long rebounds from where it had been thrown. There was no sign of Rastomaier, but he'd been outside the pods, and a lot of heavy stuff had been flung into brief, ugly motion by the impact. Including the pods themselves. Skien was out already, pushing themself towards the wall, clutching for the straps. Bartokh had his head up by then too, ruffled but unharmed. He shouted at Skien to get back in.

'It's fine!' Skien said. 'It's over. Shemp!' Calling for Rastomaier, the absent, the dead.

'Get back in, you fool!' Bartokh shrilled at him. 'The tanks!'

Tanks? I thought blankly. For a moment the habitat tanks most of us had grown up in came to my mind – the crowded, stinking mass-habitation orbitals everyone was so desperate to escape.

But he wasn't referring to those. He meant the *sample* tanks. The big armoured cylinders of Shroud atmosphere he'd been studying. Kept under twenty Earth atmospheres of pressure and filled with a mix heavy on hydrogen and methane. All around us, various shattered pieces of machinery sparked and snapped.

Ste Etienne and I both looked over at the tanks. One of them was already bloated, as though it had undergone a surprise pregnancy, straining at the seams. The impact had compromised something in its structure, and now its casing wasn't up to containing all that pent-up and eminently flammable atmosphere. Shroud was going to get us, even if it would be vicariously. There probably should have been an alarm for them, but what's one more broken thing?

I literally saw its exterior begin to split, tearing itself apart along the lines of least resistance under the incredible pressure inside. Then Ste Etienne and I were back inside the pod, the hatch slamming closed above us. At a barked command from her, a screen showed us the immediate exterior.

We saw Skien, running back for their pod. My heart leapt with utterly misplaced hope. But they didn't make it halfway before the tanks blew.

The screen glitched and I saw half a dozen fragmentary frames. The blaze, too bright for the cameras to adjust to, became something unreal, cartoonish, a bad piece of video editing. The thunderous impact against the side of the pod. Expanding, flaming atmosphere and solid chunks of debris.

The screen went dead, then sprang back to life. There was a monstrous shuddering, a sense of motion. On the screen the ravaged walls of the workshop were now sliding sideways. Thunder drummed about us, and my stomach clenched as we were spun around.

The pod was equipped with two couches, one behind and above the other. Ste Etienne was already in the lower one, her hands on the pad before her, working with a fiendish concentration. I hauled myself into the upper, the straps clasping me of their own notion and just as well, because I couldn't have secured them myself. On the screen one of the other pods spun past us in a trail of smashed lab gear and cables. I saw an articulated arm reach from past the screen's edge and snag it, then felt the answering lurch as we were dragged in its wake. We were secured to it, clasped together, whirling through space together in a zero-G waltz.

I saw the ragged edge of the workbay wall, where previously it had met the rest of the *Garveneer SP* module.

For a moment I couldn't process what I was seeing. Why was part of the ship suddenly invisible? Why was I seeing just empty space, a void without stars? And why were we being carried . . . ?

Out.

Out into space, tumbling vertiginously. We saw the ship from outside. The SP module was ruined, split like a rotten fruit, the workshop flayed open. Beyond it, the main hull of the *Garveneer*, its belly battered, receded from our view. It had cut loose from the expanding ruin of the orbital scaffold, desperately jockeying to escape any entangling debris. And receding. Each time our view passed back, they were further and further away. And, opposite them in our roll, only the dark. Not space, not stars, nothing so healthy, only the utter dark. The devouring night of Shroud.

I caught a glimpse of a panel at my elbow running through what it optimistically thought were formal drop procedures, because something in our explosive departure had registered as 'launch' to the pod. The trajectory we were on was taking us straight into the hungry gravity well of the lightless moon.

We plunged. Ste Etienne's hands, the one moving part of her, made adjustments. Around us, the emptiness began to fill with the pummelling fists of Shroud's upper atmosphere, slamming us back and forth. I squeezed my eyes shut, then the couch administered a shot of something, and the swift dark swallowed all.

1.6 DARKNESS

Strange.

Threat.

Seek.

Find . . .

Strange.

No more. Just strange.

Call.

Response.

I gather myself until I know what to do. Yet even when I learn more, I still don't know. Nothing known has been like this.

The thing in my midst: I make a ring about it.

Call.

Response.

⋆ ⋆ ⋆

I gather from further beyond. I experience what I have discovered here from all sides.

Above, I hear jagged edges. Below, pieces shift and fall, sifting into the depths. If my thoughts extend just right, they can tell me of the border layer, neither air nor water, that receives them on their way into the abyss. Its clench; the fog of it.

The ground beneath the new thing is broken. My thoughts feed me ideas of force and impact. An instant's force that I, gathered here, would be unable to duplicate. And, in the cup of its strike: the strangeness. The Stranger.

Call.

Response.

More assistance arrives and I can examine the intruder from a multiplicity of angles. My projected thoughts find it bright and round. Lying there, it sounds heavy. It tastes noisy. There is a liveliness to its dead substance. It thinks. Things move within it, like life, yet it is still as a stone. I search the shreds of my available memories, woven together, for any manner of comparison, as far back as I can recall. From this limited perspective, I know I cannot reach back far enough, and I have no past experience with which to compare this.

I have a rare moment when my mind is wide and full of knowledge, and yet I don't know what to do.

It moves. It shudders. My reaching thoughts find it dead yet full of minds. It smells of surprise.

It speaks.

INTERLUDE ONE

Let me put this into words you can understand. Life begins in darkness, where it belongs.

There is only darkness, the natural state of all that is, so why should a word even be needed for it? But there are lesser and greater darks, and colder and denser. Life began here in the utter abyss of the depths of the sea, where the hot volcanic vents spew carbon dioxide and sulphur. Where the meagre heat first catalysed reactions which accelerated mere chemistry into something self-sustaining that could develop and evolve.

These were minute flecks of insensate life, scarcely more than complex molecules within an envelope that separated *in* from *out*. Life built beside and upon that which came before it, unknowingly constructing reefs, towers and chimneys, which channelled the lifegiving heat as well as the wealth of elements from within the world. Life evolved to take account of the life it lived alongside, forming colonies which raised castles ever higher and more complex. But each mote of life was no more than that. This much, for so long, was all there was. Then, as the slow business of transformation came about, the water and the volcanic breath were alchemized by this gathering host of life. It formed into more complex configurations of elements that, in turn, created new niches for opportunistic life to fill.

In time, a very long time, these motes began to combine in more sophisticated ways than merely piling one atop another. Things of many parts arose, flat and corrugated to sift the water for ammonia and methane, and all the other by-products of simpler life. To slowly feast upon the drifting snow that was dead life, sifting it from the columns of warmed water above the hot vents. These were immobile cups and mouths and quilted things, without senses. And then, because the vents came and went, leaving only cold, dead things in their wake, creatures emerged that moved, sluggish and lethargic. Following wherever the chemistry led them. Gathering the goodness out of the water and the microscopic corpse-snow, and sequestering it in stored energy for when it came time to move again.

Others later developed forms and habits which taught them that to devour those which had fed from the rich waters was richer still, with all that wealth concentrated in one place. In devouring what was good, however, they took in what was bad too. A cocktail of poisons, metals and inconvenient compounds that would sicken them unless they could be rid of it all. And so, in sweating these stuffs from their insides, they laid down shells and plates on their outside. At first an inconvenience, these then became a defence against greater predators, and at last a support, a thing to walk within. Limbs, claws, tools formed out of all the elements that could have killed them.

But slow, still. Creeping, listless in the dark, fighting lazy battles where even the stakes of life or death could impart no urgency. Slowly altering the composition of the sea with their exhalations of life, which filtered up to change the lifeless air above too. Creating more opportunities for those forced to the edges of the water.

Life crawled indolently onto the shore, and so the second age of the world began.

PART TWO
US / THEM

2.1 LIGHT

There came a voice from the gloom, stern as the pronouncement of a deity. Snapping me back to myself, because abruptly there was someone else there, and that meant a thousand little social rules about how to comport myself were suddenly applicable.

'Stop your whimpering,' it said.

I hadn't even realized I was doing it aloud, but the voice was absolutely right, so I forced the thin, mewling sounds back down my throat. At least I wasn't alone inside the pod, which a moment ago had been a part of the great melange of terror oppressing me. Now this varnished the situation with a very fragile gloss of hope. Because, of course, when everything had blown up, Ste Etienne had been there too.

'I think,' I panted out, 'we're on Shroud.' Talking was hard. You had to breathe more often, against the drag of the gravity. The couches were very good, composed of the same tech that had supported us across the gulf between the stars, with all that crushing acceleration and deceleration. The couches, plus the tiny modifications they'd made within our bodies too, to turn humanity into an interstellar-capable species. Zero gravity or multiple gravities all became long-term survivable. Except, at heart, we were still creatures of a single G, hundreds of millions of years of evolution honing us for that very narrow and specific window. We might grow more used to the punishing conditions of Shroud, because human bodies were adaptable, and ours had been made to be more adaptable still, but it would always be hard, and right then just talking was almost too much of an effort. And I knew I was wasting that effort as Ste Etienne snapped back, 'You think?' in a tone that dripped with exasperation. Because of course we were.

The engineer was out of my immediate eyeline. I'd have to crane forwards to see her and right then I didn't feel up to that, even if I could work out how to release the straps. The hand of Shroud kept me pressed to the gel of the couch. An experience which felt like relentless acceleration, because that was my only comparator. My inner ears were kicking up a storm with it, telling me I was hurtling upwards. A bolus of nausea clenched like a fist in my solar plexus and I wouldn't even be able to jack-knife up to vomit. The wave of gorge rose with the thought, then struck some kind of barrier, leaving me constantly on the verge of throwing up for a good few minutes, yet unable to do so.

'Juna,' Ste Etienne prompted. 'Don't like your vitals.' Apparently she had access to those. I could probably have seen hers too if I'd known what commands to give or buttons to press. I wasn't able to reply, and just kept gagging, staring upwards, fighting my innards, which appeared entirely out of the control of either my conscious or autonomic systems.

Ste Etienne did something to the angle of the couches, making them revolve around one another until she emerged from below like an obstetrician. She ordered some straps retracted and looked back at me, though the effort left her pale and strained.

'We need internal cameras,' she grunted. 'Always some damn thing nobody thinks of. Just swallow it back. The cocktail you got shot full of means you won't bring anything up. Just as well, 'cos either you'd choke and die or it'd end up on me.'

I did as instructed with difficulty, but then actually felt better. With a little calm coming inside my head, the horror of everything outside it loomed larger, though. Not just the enclosing walls or the crush of the world's pull, but I seemed to feel a constant shuddering vibration from all around, and didn't know if it was from some mechanism, from outside, or just my own nerves ragging at me.

'Right,' came her voice. 'You're disengaged from the Pharma diagnostics. That's the problem. It's supposed to be reading and responding to metabolic and brain activity, but we never got the personal calibration done; go figure.'

'You sound fine,' I got out.

'I built this with me as the test subject. It's already calibrated to me. The fact you're not me is what's screwing it up. Give me . . . I'm going to reset. This is probably going to—'

I was jabbed with about a dozen needles. My vision then went blurry for a slow count of ten, as the pod's pharmacopoeia systems taught themselves who I was and started modelling my brain activity to make sure I wasn't feeling unproductive or lazy.

'Better?' she asked.

'No.'

'Well, it means things will be better from now on,' she said implacably.

I still couldn't even lift my head to look at Ste Etienne

properly, until after a further terse exchange where she showed me the basic controls for reconfiguring the couch, all conveniently under my left hand. There was voice activation too, but she said I should learn the fingertip controls because she didn't want to hear me arguing with the furniture all the time.

Using this advanced tutelage, I had the couch prop my head up, like some phantom orderly plumping my pillows. I could finally see her properly, Ste Etienne, Special Projects engineer. Her neuroplot notes mostly focused on the fact she was very good but needed to be allowed a lot of leeway if she wasn't going to spend her energies frustrating her superiors. Exactly the sort of person Bartokh had needed me to manage.

'Oh God,' I sputtered. 'The others—'

'Don't—'

'Skien, they—'

'Your heart rate's spiking. Can't afford that right now,' Ste Etienne said, brutally calm about it all. 'There are other things—'

'Did they make it? Skien?' In my mind's eye the geologist was kicking back towards the safety of the third pod. Maybe they'd made it. Maybe, in the chaos of that explosion, they'd squeaked inside and slammed the hatch shut. They could be out there – alive.

Ste Etienne made a long-drawn-out sound of frustration. 'Okay, look. Skien didn't make it.'

'But they were running—'

'Not in time. They're dead. Rastomaier's dead too. Process that, can you? Get it out of the way. Bartokh I'm not sure of. Still trying to bring stuff back online after impact. Don't even know if he's still grappled to us. Tolerances say it's possible, and some of the drop data suggests his pod was chuting and retrogassing with ours at least some of the way.' She was fighting the words out, but she was better at it than me, and growing used to working against the press. The actual words, though, meant next to nothing to me right then, until she said, 'Get your head together enough to help, that would be grand.'

Something scraped distantly, as though the sound – the vibration really – was coming to us from the far end of a long tunnel. Long and infinitely dark. I remembered that exploratory rap, from before Ste Etienne's company monopolized my attention.

A quiet descended, grim and thoughtful, until I broke it by saying, 'Something's out there.' I was still trying to manage a regular cycle of drawing and releasing breath, but it was proving difficult. Each inhalation had to be micromanaged, like my autonomic systems had just clocked off.

'Bartokh,' Ste Etienne said. 'I hope.'

'And if not?'

'Don't know.'

'Did we . . .' There were more reasons than physics preventing me from keeping my breathing steady. 'Did we come down anywhere near where Drone Forty-Eight was?'

Ste Etienne was facing away from me again, because anything else would have been too much effort. Her hands had been busy on the controls set beside her couch, her eyes on a series of small screens projected by the panels in front of her. Now she stopped moving entirely, just looking at nothing. In the unnervingly long pause, something else scraped almost gently along the outside of the pod's hull.

'I don't know where we are,' Ste Etienne admitted. 'There is literally zero data I have that might tell me. Odds are damn small, 'cos Shroud is big. Don't know how the wind system figures, though. It sure as hell grabbed us at one point. Maybe there's a big rubbish tip of crashed drones, all on one place. In which case, yes. But we both agreed . . .' She wheezed; too many words, too little breath. 'Agreed the goddamn spanner in the footage was an *artefact* of the clean-up process. I'm sticking to that. Not going to believe anything on Shroud has . . . has evolved a modern disassembly toolkit, okay?'

'Okay.' I was at last finding a laboured rhythm of lungs and ribs that seemed to be self-perpetuating. The gravity was becoming just one more part of the grotesque situation that my augmented body had been designed to cope with. In the wake of that, I did my best to start taking stock. I even invented more problems, in fact. I told myself that our wage-worth had massively decreased. It would be back to the habitat tanks, back on the shelf, they'd never trust any of us with valuable work again. Sharles Advent would personally express his disapproval of us for damaging his ship. And I'd take it. Every punishment, demotion, censure and black mark on my record. I'd kiss Umbar's boots. Because at that point I wouldn't be on Shroud any more, and not being on Shroud had just become my life goal.

'What can I do,' I asked Ste Etienne, 'to help?'

Most of what I could do, in the next couple of minutes, was run a range of checks that turned various indicators from red to amber or green, the perennial colour code every kid from the habitat tanks learned before their first words. A few of these I understood, others related to processes entirely invisible to me. As I did it, I kept trying to ask about everything else: the planet outside, the noises, Bartokh, the *Garveneer*. Ste Etienne shut me down each time.

'Priorities,' she said. 'If what we have in here doesn't work, the rest won't matter.' Just like it was when we were in space, the hierarchy of needs was weirdly inverted. The absolute baseline required for life was that a vast amount of highly technical systems functioned, because without those it was either hard vacuum or the hostile conditions of an alien planet.

We managed to establish enough green amongst the rest for Ste Etienne to declare herself satisfied. Even as she did, another metal shudder went through us, and I fancied I felt the whole pod tilt, very slightly. It left me wondering if Ste Etienne's laser focus on the internal systems had been entirely wise. No point making sure your life support is working if an alien monster with a multitool was about to open the whole show up. Or if the considerable weight of the pod was about to crash through whatever it was sitting on. I remembered the weird interwoven briars that had been the substrate in the drone recordings, and the unknowable depths beneath them.

'Right then.' Ste Etienne still sounded calm, in a situation where calmness was surely insanity, and ranting and shrieking should have been the norm. It was only her presence that stopped me howling like a wounded animal, honestly. Only the discovery that sheer social embarrassment was apparently a stronger driver for me than fear of death.

'Reactor's green,' Ste Etienne counted off. 'Air scrubbers, green. Water recycler, green. Rations . . . Guess after a couple months we'll need to eat each other. Actual wear and tear, degeneration of the whole shitshow from pressure, Gs and alkalinity . . . I mean, this is basically the field test, right? We only had cores come back from the drones – never saw how their casings did. So available data is zero. Right, let's take a look outside, shall we?' A beat. 'Okay, you already did.' And there it was, something I'd already realized that she hadn't. The dark screens hanging in the air around us were not just shut down, they were what our cameras were actually seeing. It wasn't just any old nothing. It was the authentic alien nothing of Shroud, live from outside.

The engineer sucked at her teeth, and for a moment I thought she was finally unnerved. 'Lamps dead then,' she said. 'Let's get them fixed up. How about—' There was an appalling blast of static. Not even static, a cacophony of squealing, howling and roaring. The voice of Shroud, just as we'd heard it when we first settled into orbit around the moon. For a moment it was in the pod with us, shaking every surface, shivering the gel of the couches, rattling the teeth in our heads. Then it was gone.

'Damn,' said Ste Etienne in the sudden, shocking silence. 'Forgot.'

'What was . . . ?'

'Radio. EM receiver. Told it to find signals. Standard procedure after a crash, right?' She did sound ever so slightly rattled now. 'That's . . . outside. Brace yourself. I'm going to test audio.'

I was expecting the sound our radio receiver produced and the actual sounds from outside to be basically the same. Just a wall of signal so compressed it became noise. It was certainly impenetrable, and it was just as bad, but I was becoming a connoisseur of the horrible by now. The actual sound of Shroud, even with the volume dialled all the way down for human comfort, was like every single percussion instrument ever invented being played by frenetic maniacs, all at once. A thunderous drumming, rattling and ringing, and someone running the sticks up and down a glockenspiel with frenzied energy.

When Ste Etienne shut it off, I realized we'd solved the mystery of the constant background vibration. It wasn't some labouring system of the pod, but the surrounding soundscape, deadened by all the layers of hull until it was nothing but that background thunder, going on and on and on, because nothing on Shroud needed to take a breath.

Another contact sounded from the blind dark out there. A solid object rapping our outer hull and grating across it, exploring its contours. The impact came to us faintly but distinctly, distinguished from the rest of the background chorus because it was something inarguably connecting with the hull from outside.

'The briars,' I said. 'Broken ends of them. Or something.'

Ste Etienne grunted, which meant she was working. She turned on the radio receiver again and the sound was . . . not better, really, but different. More refined. It shifted from shrilling to squealing to staccato yammering to roaring. Selections of patterned nonsense from the asylum radio station.

'I've got some algorithms hunting for patterns,' she told me. 'Except it's all patterns. So still trying to refine them.'

'For what?'

'In case . . .' I don't know if she even had a back end for that sentence. If she actually had a goal, or was just doing a thing because it was there and could be done and would take her mind off the overarching hopelessness of our situation. But abruptly it was moot, because the algorithms shouldered all that idiot noise aside and narrowed in on a single frequency, one signal. Static chewed about the edges of it, and a choir of damned souls was constantly playing beneath it, trying to prise their way in, but it was a voice. A faint and tattered voice, coming to us as if from the other side of the grave.

'Hello? Anyone receiving me? Hello? Other pod? Mai? Hello?'

A jolt went through me. A spasm of absurd relief. Bartokh. It showed how desperate the situation was that hearing Bartokh's voice made anything better. Yet his voice was coming through clear enough to recognize, even though the fluting, whistling scour of Shroud's radioscape constantly threatened to overwhelm it.

'Confirmed receiving you,' Ste Etienne said, and then something banged particularly loudly off our hull and we lost Bartokh's signal entirely. Cue panicked minutes of Ste Etienne doing most of the work and me running diagnostics on our systems, while the howling all-wavelengths Armageddon that was Shroud's radioscape came and went.

Something very solid shook us, and we lurched.

Both of us stopped. I whimpered and this time Ste Etienne didn't tell me not to.

'Hello? Are you there?' Bartokh again, somehow. Clearer now, in fact. Still as though he was five cells down the corridor in the Home for the Noisily Insane, but we could hear him.

'Connection re-established,' Ste Etienne reported. 'We are receiving you. Don't go anywhere.'

A faint, dry chuckle came through from him, in which I could hear the wheeze of lungs under Shroud's heavy hand.

I understood, then. Or at least I understood some of it, and Ste Etienne filled me in on the rest. She'd grappled us together, and whatever fixings the pods came with had proved strong enough to hold the two as one all the way down. Our stacked deceleration measures had deployed in sync too, keeping us together without severing the connections. We'd broken apart only on impact, and the impact itself had been relatively gentle, as demonstrated by the fact we were still whole and alive. There had been parachutes and gas thrusters, all triggered automatically over the long and slowing fall we'd been knocked out into. Plus we'd come down on a surface with enough give in it to cushion us. Honestly, Mai's engineering specs deserved three quarters of a medal, and the unexpected cooperation of Shroud's landing pad the other twenty-five per cent.

Then Bartokh's pod had rolled away, slightly. Literally just that, after we landed and the pods had, apparently, decided that was their cue to disengage. And even a short distance had forced our signal-detection algorithms to work overtime to pick up his transmission. We were talking through a physical link now. Between them, he and Ste Etienne had blindly groped about until the two pods could shake hands again, providing a signal through the contact. It was still tenuous, but he had us and we had him. We were three.

Three out of six from the Special Projects team. Four, maybe, because there was every reason to hope that Jerennian had survived, up in his bunk, away from the ground zero of the workbay.

'You have no idea,' came Bartokh's corroded voice, 'how glad I am. To hear you.'

'Take over,' Ste Etienne told me, and then ran me briefly

through which couch-side controls governed my access to the channel. 'Still have a queue of stuff to fix. Talking's your department.'

I supposed she was right, honestly. I was always the person who liaised between the Director and the rest of the crew, even though that crew was now just one woman.

We lost him for a moment after that, to general panic. Whatever we were resting upon shifted, and the physical connection between the two pods slipped. Even with us positioned right next to each other, all Ste Etienne could do was grope in the dark with one of the waldo arms fixed to our pod's exterior. One of the *surviving* arms, she said, because a couple were returning red lights and might have been sheared off or buckled into uselessness.

She talked very blithely about all this. I was only just understanding how lucky we had been to come down in one piece at all.

The rattle, bang and scrape of missed mechanical clutches came distantly to us as both pods' arms fumbled about for each other. At last we restored connection, and Bartokh's etched tones came back to us, against a painful whistle of feedback.

'I don't know what our situation is. I can't pick up anything from outside except all that noise,' he told us. 'Juna, I need Mai to signal the ship for rescue. We just have to hope they're in a position to act on it.'

I opened my mouth, then looked to Ste Etienne, who just hunched her shoulders, not even sparing the effort to glimpse back at me.

'I don't think that's going to happen, Oswerry,' I said weakly.

'If your transmitters are damaged then have Mai guide me through what I need to do.'

'Oswerry . . .' I bit my lip. It seemed so enormously obvious to me, but here was Bartokh, no fool he, still telling me to do it. 'Mai, I . . .'

'Ask him if he took a hit to the head or something,' she said uncharitably.

'I didn't catch that. Repeat please?' he said.

'Oswerry. Director. We can't . . . there's no way we can signal the ship. Because of the ambient radiosphere. The noise.'

A pause, while Ste Etienne continued to work, and then Bartokh replied, 'I had assumed these pods were fitted with . . . something. High-power transmitters. Some sort of targeted light beam signal or . . . something.'

'Atmosphere's too occluded to pass a light beam through,' Ste Etienne said tiredly. 'Ain't going to happen, Director. We cannot reach the ship. Ship cannot reach us.'

'Can we . . . go high? High enough? The interference lessens with altitude if I recall?'

'Ask him where the nearest mountain is,' Ste Etienne said. When I didn't, she went on, 'We got almost nowhere with topographical flyovers. Just too much competing signal for the drones' instruments. Plus there's the little matter of having zero idea where we've landed.'

'What?' I asked, without even needing Bartokh's prompt. 'We must have some idea. You said . . . maybe where the drones went . . . ?' That wasn't what she'd said. It was just about the opposite of it in fact. But it was such a basic thing, to be able to find out your own location. It's amazing what you take for granted.

'How can we know where we are?' Ste Etienne asked.

Normally we'd rely on triangulation with satellites, except here we couldn't receive any kind of signal from orbit. 'We . . . I mean there's a magnetic field . . .'

'Which doesn't tell us where we are,' Ste Etienne pointed out. 'And the local yammer interferes with it. I can just about make out which way the magnetic pole is, but that still doesn't help us. Because what, exactly . . .' The punchiness ebbed from her voice, into a more thoughtful tone. 'Maybe there *is* somewhere we could aim for, but I don't know how we'd find it. We'd need more data points.' She let out a long breath. 'Okay. I think I've restored the lamps, or some of them. Want to see?'

Just before she turned them on, I realized I didn't. I understood, in that last moment of blessed ignorance, that ignorance was surely the best possible state for us. And yes, we did actually need to know the situation outside in order to achieve anything at all. But it could only be obstacles and problems. It could only make things worse for us, even if we were only able to save ourselves by crawling over whatever it presented us with. And even if just wallowing here in our blindness was the sure, slow road to death, there was nothing that could be illuminated outside that would bring us joy.

The lamps flashed on, searing the dark screens into chiaroscuro patterns of stark light and impenetrable shade. And it was worse than I'd feared.

I probably whimpered again, but I could hardly be blamed for it. In the background, Bartokh's tremulous queries continued, but for a long while neither Ste Etienne nor I had the ability to put anything into words.

One screen did show Bartokh's pod. It was at an angle, one leg projecting awkwardly up, caught in a cup of broken brambles and caressed by their feeding fans. The network of branches was finer here, a different species than we'd seen before. They had already started to regrow around him, a basketwork cage that his vehicle would surely never escape from. We were most likely in the same situation.

I didn't care about that right then, though. We had other worries.

They were all around us. A dozen, twenty maybe. The beasts from Forty-Eight's recording, the ones which had seemed to manifest a disassembly toolkit in those final heavily restored frames. Despite the fact we now had a real, good view of them, their weird construction still baffled the eye. They were all woven and interlocking, with open-work sections sliding past one another. Occasionally the light would catch them at a precise angle, pierce through the hollows of them, and we'd see something leathery and flexible within. Every surface of them, stark in the unforgiving light, was heavily textured: whorled, pitted and ridged with teeth, as well as peppered with holes. And each hole held a kind of flexible collar around a fine, gleaming hair. The whole thing shivered and shimmered constantly, scattering muted rainbows like handfuls of ephemeral jewels as it sliced our white light into fragments. Except the oppressive, thick air sludged the rainbows into sickly, curdled shades.

They were in constant motion. Not a rapid spider-patter but a deliberate stalking, that was somehow worse. Each limb – that weird, offset plan which alternated left and right down the lengths of their bodies – extended with a dreadful premeditation, reaching for purchase. We could see no spanners, but a kind of hinged claw arrangement at their tips, which clasped shut around a stem seemingly automatically, releasing when the limb was withdrawn again – an ingenious mechanism. And now we could see those limbs up close, with barely a gust of static in the way . . .

'Damn me,' said Ste Etienne. 'What the actual hell?'

The nearest creature to one of our cameras took a step forward. The light was well positioned to show us the series of moving parts that made up its limb, and how each fed back into the body of the creature, like a puppet moved with rods. Each leg was just an empty armature. The beast itself did not inhabit it, but lurked within the main framework, a living creature inside its complex articulated shell.

Fans or gills erupted from a series of openings we hadn't even noticed, briefly doubling the creature's apparent size. They trembled for a moment, feeding or breathing or making a threatening display. The faint thunder of Shroud's soundscape vibrating our outer hull seemed to change slightly, as if it were roaring at us. Or was that my imagination?

Bartokh's pod, I saw, was at far arm's length from us. No wonder establishing a physical connection had been so hard. The artificial arms of both pods were at full extension, battered and dust-scoured in the light. Which meant that all the scraping and banging we'd heard since we woke hadn't been Bartokh after all, as I'd been comfortably assuming from the moment we heard his voice. The things had been patiently investigating our pod.

'What? Talk to me. What is it?' Bartokh's demands, ongoing but ignored since the lamps went on. And they went unanswered still, because we weren't done staring.

They were shaped like maggots, save for those alternating legs. A tapering front end, going into a fat middle, then a tapering rear, with no distinct head or tail, just those repeating segments which swelled and diminished along their length. They were perhaps nine metres, I estimated. And maybe three or four across the thickest part of the body. How big the actual living portion was, I had no way of deciphering. The majority of the creature's bulk seemed to be the complex interwoven struts of its casing.

As for what they were doing . . . With no eyes, it was impossible to say for sure what they were focusing on, but their front ends were directed at the two pods more than anything else. I remembered that spiral of diminishing limbs as the segments grew smaller and smaller, and they kept reaching out towards us. I watched them touch the curve of Bartokh's hull, tapping and scraping at it in a way the human mind could read thoughtful consideration into. Like someone with a hammer testing a priceless vase. Sometimes they used those little fine arms at their very snout, which I guessed were at least as delicate as human fingers, perhaps even finer. At other times they flicked out longer, nastier-looking limbs from within their bodies: thick-sparred prying levers, with hooked ends that spoke sheer brutality to my eyes. And yet they didn't follow through, always slowing the strike so we just heard a dull impact through all the layers of hull.

Keeping track of them was impossible. Individuals kept giving way to others. We didn't observe any of them snap at one another to assert dominance, or shove and shunt each other aside to grab a taste of the strange new arrivals in their world. They were constantly moving past us, and each other, investigating. Not locking limbs, not even physically brushing those metal hairs. I was the one who noticed that they maintained a very specific distance, always at least a thumb's width of clear air between their ghostly filament pelts.

'Director. Oswerry,' I said, with what I felt was admirable self-possession, 'we are surrounded by Shroud life. Macrofauna. Literally surrounded.' I took a deep breath, the act of having to report to Bartokh serving to give me focus and keep me calm. 'We are resting on more of that bramble-style terrain, just latticework over . . . over darkness.' A little wobble in my voice there, but forgivable. 'Your pod is caught in it, having broken partway through.' Maybe it was briars all the way down, or maybe this was just a thin layer suspended over an abyss. The infinite darkness of a freezing and yet unfrozen sea. And what sea on this world would be free of even more terrible monsters?

'I see.' Bartokh was also playing the self-possession game. 'And would you be able to free me, do you think?' 'Would you' as a qualifier, because such efforts could probably only be achieved in a world free of monsters.

'We have . . .' I checked the green lights. 'Three working arms, I think, and one with limited functionality. With your help we might be able to pull you onto unbroken ground.' Another heavy bang sounded on the outside as one of the natives tried a new mock-attack, or warning, or attempt to test our hull's integrity.

'How are you with arms?' Ste Etienne asked me.

'I . . . what?'

'If I shift us closer to his pod, can you actually do what you said, about pulling him free? Or was that you speaking for me?'

I felt my face flush. 'I can work the arms, yes. I had the training.'

Her grunt communicated the world of difference between that and actual experience, but apparently she'd take what she could get. A moment later I let out a squeak of terror when our pod lurched. The substrate had given way beneath us! The monsters had grappled us! But it was just Ste Etienne getting the pod's legs working, pushing blindly against whatever they could find purchase on, lurching, leaning towards Bartokh, then steadying. We went alarmingly sideways, then our gyroscopes adjusted, finding the vertical the way the pods were supposed to, as the main body moved within the ring the legs were fixed into. Bartokh's vehicle looked skewed to us, because the legs were all up on one side of it, but the interior would hopefully have rotated to keep him the right way up from his perspective. Hopefully.

The monsters had gone still at this display. One shivered, and then another, throwing off those spectra of corrupted colours. I heard another solid rap on our outside, and then another. One screen showed the corkscrew front end of a Shroudbeast fumbling around where the camera was set, a forest of arms pistoning in and out as it scrabbled at us.

'Mai,' I said.

'I see it. I'm going to try and give us some breathing space. I hope,' she said tightly. 'Ready with the arms?'

I nodded, then registered she couldn't see and said, 'Ready.'

'What's going on?' Bartokh demanded. 'Someone talk to me.'

'Going in three—' Ste Etienne said.

'Wait, going what?'

'Two—'

'Mai, what are you—'

'And go.'

The images on the screens fuzzed into chaos for a second, and when they were restored the monsters had . . . reacted. Gone very still again. I almost forgot about my part of the deal, but then activated each arm in turn, reaching over from the top of our pod and latching onto Bartokh's. I was trying to work out on the fly how to get the best purchase to lever him from the fractured grip of the vines. I focused only on that element of the screens, ignoring the way every monster seemed to be pointed right at us. Not Bartokh, but us.

'Well, that made some impression,' Ste Etienne said.

'What did you do?'

'Big EM pulse, all channels. Shouted at them. Or maybe made ourselves look super-big, depending on how their senses work. Was hoping they'd spook, or something . . . Just . . . three big pulses, *boom boom boom*. Look how loud and scary we are.'

'You got their attention.'

'I did that,' she agreed. 'They haven't eaten us.'

'Yet.' We tilted as my efforts to pull Bartokh out shunted us. 'Can you give me a better footing?'

Ste Etienne grunted, and we rocked and shifted. The monsters were moving again, and I thought they were more agitated now.

'I think they're angry,' I said, hearing my own voice shake. A tide of panic was rising in me, because I couldn't shift Bartokh's pod and the periphery of my screen was filled with all that stalking, alien motion. The reaching and retracting of their bizarre lever-legs.

They went still once more, just for a moment. Ste Etienne

had let loose another pulse. But it didn't halt them for even the few seconds the first one had, only increased their animation. They weaved around the far side of Bartokh's pod. The background buzz of sound ramped up until I could feel it in the roots of my teeth.

'That was the loudest we could shout,' Ste Etienne said hollowly.

'I think they can shout louder.'

'I don't seem to be moving,' Bartokh's static-eroded voice put in helpfully. 'Can you make it to me?'

'I'm still trying to establish proper purchase.' I was moving the arms, but the rounded hull of the pods didn't offer many grip-points and I'd already tried the most promising of them. 'Can you get your legs moving? If you push yourself out with them, I can try and help . . .'

'You're going to have to guide me through where the controls are,' Bartokh said.

'No, look,' Ste Etienne broke in. 'I'll slave your pod to ours. I can walk you, with help from your pod's automatic systems. Okay?'

'Just do what you have to.' Bartokh's voice was barely audible, waves of interference breaking into the transmission through the inadequate shielding of the physical link. 'But if you could manage it sooner rather than later, I'd appreciate it.'

'Oswerry,' I said, suddenly breathless. 'Keep talking.'

'I'm . . . fine, Juna. I did sustain some injury coming down but not a head wound or anything. You don't need to be concerned about me personally. The, ah, the pod's shot me with painkillers and stimulants. I'm quite lucid.' The awkward ramble of a man forced to articulate more than he actually had to say. And all the while behind it was that intermittent pulse of static. Ste Etienne had noticed it too, now. She actually shifted in her couch to lock eyes with me. When Bartokh trailed to a halt, she held the channel open, and turned the open mic to the outside world. We both listened.

Boom boom boom. Three distinct electromagnetic pulses, the gap between the second and third slightly longer than between the first and second. A pause, and then it was repeated.

'You said—'

'I sent the signal exactly that way, yes,' Ste Etienne agreed. ''Cos my finger stuttered when I did it. I'm going to try four next.'

For a long while there was nothing, or rather there was just everything. The great all-bands chaos of Shroud screaming at itself. Bartokh querulously demanded to know why he wasn't free yet, but neither Ste Etienne nor I had any ears for him.

Then we picked it out from the chorus. Four pulses back. More, four pulses centring around the frequencies that Ste Etienne had put most of our signal into, and following exactly the same slightly irregular pattern. An echo. A reply.

2.2 LIGHT

Ste Etienne tried again. Three pulses, on a narrower band of frequencies this time, the ones they used themselves. Listening in, even just to that band, we could still hear a howling cacophony, slightly strangled by our focus but fierce and wild and incomprehensible. Until the things answered, their pulse coming back strong enough to blast through the static, as they crowded around us. I was suddenly jolted back into remembering I'd been the one studying the surface radiosphere when I was back on the ship. Using the couch-side controls, I pulled up a screen and hoped to God that Ste Etienne had already loaded up at least some signal analysis routines into the pods. That we weren't just an ambulatory shell with no research capability whatsoever. Thankfully most of my toolkit popped up, buried down a nest of folders. I took the signals the aliens – in my head they were the Shrouded, which sounded sinister as hell – were making and ran some basic inspection of them, looking for details that the human eye couldn't pick out from all the static. There was a lot of complexity in that signal; a variation of volume and pattern within the three that was almost-repeated each time. Too closely for mere noise or coincidence. My heart was in my mouth as I fed it back to Ste Etienne.

'What now?'

'Use this algorithm in your next pulse, please. To add in some pattern coding.' It was hard to know if it was the gravity or suspense which made it hard to breathe after she'd sent it.

The attitude of the creatures changed distinctly. They had never become one hundred per cent still, but the first set of pulses had monopolized their attention, drawing their busy front ends towards the pods in a way that told me at least some part of their sensory apparatus was directional, eyeless though they were. With this new set of signals they seemed to lose interest, however. Individuals remained pointing at the pods, but others turned away, trading positions and orientations constantly across the group.

'So what was that?' Ste Etienne asked.

'There's a lot of traffic on the band they're using. Just from . . . everything else out there,' I said. It felt weird to be explaining something to her for a change.

'I noticed that, yeah.'

'So if you're using a signal for anything, either comms or radar or whatever, and you need to beat the competition, you can go louder, you can go higher or lower, or you can go complicated. So you recognize your own signals coming back at you. Like a code. Like the teeth of a key in a lock.'

'Like a . . .' For a moment the analogy left her blank, because this was a historical reference if there ever was one, but then she got it. 'Well, it did something. What did it do?'

I worried for a moment I'd somehow ruined everything. Their focus on us had been unnerving but it had been something. Any interpretation I made of it would be nonsensical anthropomorphism, far more so than if I had projected human intent onto a cat, lizard, or beetle even. Except . . .

The simple fact they had a front and a back began to narrow down the options, introducing a potential convergence of the broadest strokes of behaviour.

'I mean . . .' I began, about to go down a hole of speculation that would have seen me stripped of my scientific credentials if anyone had been here to do so. Ste Etienne wasn't even looking at me but I couldn't go on. It was magical thinking. Helpful aliens recognizing the star-travellers in need. How utopian would that be!

'Once again I am forced to ask for some sort of clarity,' said Bartokh testily, and then he actually screamed, just a brief bark of terror before his signal cut off. I was instantly panicked, even though I could see his pod in our lamps and nothing visible was happening to it. Maybe one of the monsters had bored in from the far side, or the bramble substrate had cracked it open from beneath?

Ste Etienne's shoulders were shaking. 'Get him back!' I told her. 'Link to his pod! Find out what's wrong!'

'I'm linked to his pod,' she said, her voice shaking a little too. 'I just turned on his lamps and screens.' I realized then she was laughing, or at least trying not to laugh.

'Oswerry . . . ?' I asked warily.

'I . . . I'm here. Yes. That was a shock. Have Mai warn me before she pulls anything else like that.' Talking like we were in a meeting, director and assistant, even though Ste Etienne was right there listening. Chuckling at him. The first laugh on Shroud.

'What's the situation?' Bartokh pressed. 'You were silent for a while there.'

I found the link to his pod and had my routines compose a summary of our discoveries for him. The mental image of him reading over them, forehead furrowed over his wispy eyebrows, was absurdly comforting. We were establishing a veneer of normality.

'Well,' he said at last. 'This is unexpected.' The understatement of the galactic year. 'Have you tried four pulses?'

The static had risen as he spoke, even though we were trying to use wavelengths that the surrounding monsters didn't seem to favour. I had to ask him to repeat himself, and then the words didn't seem to make sense.

'Or two,' he clarified, in a lull. 'But four would be better. Build towards a greater complexity.'

The monsters outside became restless suddenly. Or restless-adjacent, given their weirdly deliberate, stilting motions. I wondered if we'd started to bore them.

'We did four already,' Ste Etienne agreed. 'Let's try it again with frills.' Meaning our mimicked variations on a theme.

Almost immediately we received a response. Four beats to the same pattern. The slight irregularities of a manually keyed signal, replicated perfectly in an alien response.

Ste Etienne swore, and so did Bartokh when I told him.

'Try some maths. The Fibonacci sequence. That's basic stuff. Oh God, this is huge. This is incredible.' There was a crowing tone to his voice which suggested he was picturing someone giving him a promotion and an award, as well as a big hike of wage-worth for this. Conveniently overlooking the fact that, unless Shroud had independently evolved an academic credit program, there was nobody around to care.

The static on the channel we were using to talk started gusting more fiercely, as though there was an electromagnetic storm on the way, but we could still just about make him out. Ste Etienne was sending even before he'd finished speaking. One pulse, one pulse, two pulses, three pulses, five, eight, thirteen . . .

The correct response for the aliens to fit into the pause would have been twenty-one, the sum of the previous two sequences. Instead, they just echoed each transmission back to us, one, one, two, three, all the way to thirteen. And when we went silent, they didn't leap in to extend the sequence. Fibonacci would have despaired.

'We know they can count to thirteen, I suppose . . .' I said.

'Try it again,' Bartokh insisted, though we could barely hear him. Only the rhythm of his speech came through, the human rise and stress that gave us a best-guess at his words even without most of the consonants.

'Later. Time enough for a maths lesson when we're actually mobile,' Ste Etienne decided. 'Let's pull the old coot out of there.' Our pod rocked and swayed as we both fought with the grappling arms again, scrabbling for purchase on the awkward, piecemeal ground beneath us. I distinctly felt something we were pressing against crack and splinter. A chill washed over me that was, honestly, entirely irrational. Yes, we could end up as stuck as Bartokh, but in the cosmic scheme of things did it really matter? Because we were just as stuck on Shroud, even if we were able to roam the whole planet for however long our pods kept us alive.

I had the couch shoot me with a mood stabilizer. Standard working practice when despair was creeping in. Yes, the situation was particularly extreme, but it wasn't the first time.

Ste Etienne cursed, and we slid sideways, tilting us over before the gyroscopes of the pod compensated. 'Damn but he's in there solid,' she said. 'I think one of the legs is hooked.'

'Are these pods equipped with the ability to gnaw their

own legs off, if they're in a trap?' Bartokh said, with a valiant attempt at dry humour that amazed me. I had seen him under the hammer of budgets and deadlines and he usually tended to export his stress onto me. Right now, though, when it was his life on the line, he was eminently calm. 'You may have to leave me,' he went on, and I wondered how much of that calm was chemical, shot into him from his couch. 'If, that is, you have anywhere to go.'

'Maybe we do,' said Ste Etienne. She'd said as much before, but I'd thought it was empty words to keep me going. Apparently there *was* a plan, though. My heart leapt. I found that so long as Mai Ste Etienne had an idea, that made everything better. I very much didn't want her to explain it to me, however. Actually knowing what was doubtless a wretched piece of cobbled-together nonsense would only kill the fragile mood of optimism.

Then we lost Bartokh entirely because something gave under us and we rocked backwards, dropping what felt like two metres. Ste Etienne kicked into motion, wrestling with the controls, and for a moment the screens were a mess of criss-crossing briars and flailing metal limbs as our goldfish-bowl world rotated vertiginously. After which we were up again, clawing our way back to the tissue-thin layer that passed for the surface.

I heard Ste Etienne cry out in alarm, but for a long second I couldn't even make myself look at what she'd seen. One of the cameras had been pointing down, partway through that scramble. I had seen only darkness, crossed with a tenuous lacework of briars. No land, not even water. Just void.

Then I checked a different camera and saw the Shrouded were swarming Bartokh's pod.

We'd climbed back up around eight metres distant from him – too far to re-establish a link and our groping light only dimly reached him. Whatever cries, whatever noble oration might be coming from the director, we couldn't hear it. The Shrouded's wickerwork bodies had eclipsed his pod entirely, and the murk of the air obscured what they were doing to him. I caught the patient, murderous arch of their extending limbs, saw their legs reach and clasp for purchase, and their entire segmented bodies tense and flex with effort. In my mind they were tearing open the casing of the pod layer by layer. We rocked and jolted as Ste Etienne tried to drag us over there, fighting for purchase with the pod's broad, flexible feet. She'd done her best, designing them for the terrain we'd seen, but there was a howling gulf between theory and practice.

Then the Shrouded ebbed back. Despite the huge size of them and their stop-motion movement, I was reminded of rats. Humanity's eternal co-passengers and experimental subjects. A horde of hungry rats leaving nothing but the bones. Leaving . . .

Leaving Bartokh's pod upright, undamaged, levered clear of the snarl. Placed neatly beside the skeletal crater he'd made when he came down beside us.

Ste Etienne piloted us back to his side, lurching step by step. Neither of us had any words. When we clasped hands with Bartokh again – more quickly now, because we could all see what we were doing – he was similarly dumbstruck.

They had understood. We were all thinking it, but nobody wanted to put their head out of the scientific trench to say it. They'd seen what we were trying to do and grasped it. Which implied a world of commonality, surely. Not just an understanding of us as distinct entities, despite our alien nature, but familiarity with the concepts of aiding others. A social species. Intelligence.

'Maybe they can help us,' I said at last, because it wasn't as if anyone would be sending me back into hibernation for it, or back to the habitat tanks.

We rocked again. 'Um,' Ste Etienne said. I thought the substrate was giving way again, but then another couple of impacts clanged dully on the exterior and I saw that we were being pushed. The creatures were shoving us, and Bartokh too. Herding us.

'Maybe we don't want them to help us actually,' I decided, although they were all around us now, and just carrying us off seemed entirely possible, if they could take the weight. The pods, of course, weighed a great deal under Shroud gravity, and even if the Shrouded were strong enough, the ground beneath us probably wasn't.

Bartokh was saying something but the tide of interference entirely eclipsed his words, and a moment later we found out why.

There was a brisk rattle, just a distant patter on the outer hull, like rain. Brief, broken motes danced in the cameras, catching in the light of our lamps like dying fireflies. Shattered pieces of bramble were showering down on us from above. I hadn't considered that, after airbraking as well as venting gas thrusters as much as we did, the pods had still crashed hard into the surface of Shroud. Sunless as this world's sky was, I hadn't realized that we were already buried. We had smashed through a significant layer of the briary substrate, perhaps many layers, before coming to rest here. If there had been rock where we came down, likely neither we nor the pods would have survived.

Knowing how devoid of light the heaven of Shroud was, I had never even thought about looking up. And I discovered I wasn't the only one. As more and larger pieces of shattered stem now clattered down on us, the surrounding aliens were going into a frenzy. Ste Etienne was very still when I told her to show us what was going on.

Then I understood, as I saw the monsters jutting their forward ends towards the sky. Flaring their whorled fans of limbs. Surely this wasn't a visual threat, because we off-world visitors possessed the only sense organs that could appreciate it. And our eyes were useless, since the pods had no cameras that looked *up*.

Ste Etienne was hissing to herself, fighting with her couch-side controls. 'Drone,' she managed to say. 'Eyes in the sky.' But she couldn't get it to launch. Like every other system, it had suffered in the descent and landing, and it had been a long way down her priorities list until ten seconds ago. I could hear Bartokh demanding to know what was going on, again, but he was down the list too. We had no answers for him anyway.

Abruptly, something was there, shockingly sudden, as though it had been placed in a gap between seconds that none of us actually experienced. A pillar, gleaming like steel in the light, thrust down between two of the monsters. It was worked into a spiral, like a narwhal's horn or a braided bundle of cables. In another eyeblink it was gone.

Then we were on the move, the monsters jostling us along. We tipped and lurched dangerously, Ste Etienne struggling with the pod's legs until she was able to position them back below us and fumbling at the substrate. In that moment I caught a glimpse of *it*. We tilted far enough that our side cameras were pointing straight up for just a couple of terrifying seconds.

The furthest range of our lamps caught a network of grey-white interlocking stems, jagged and broken, and past them something moved. It was shaped like a broad, cupped leaf, as best I could see. A leaf rotted down to the tracery of its veins, skeletal. There was a thing visible within this frame, like an anatomized bat, all fluttering membranes and twisting tendrils. In the instant I saw it, it was bunching and flexing, anchoring itself within its frame for purchase so that it could manipulate great stilting limbs, which extended far beyond the reach of our lamps. It appeared to be a vast spider of a thing, more than fifteen metres long in the body, but surely many times that in leg-span. A perfect predator to pick its way over the porous forest of briars and . . .

There was a folded-up process held within the concave underside of its woven frame. Something of the beak, something of the industrial piston. Then we righted ourselves and I lost sight of it. We were underway by then, and I had a sudden, horrified fear that Bartokh would just be left behind, cut off and complaining, to then become prey. Except Ste Etienne had slaved his pod to ours, and the leg algorithms were doing what they could to power us along on the spectacularly uneven ground. Around us, the throng of alien monsters were swarming. It almost seemed as if they were forming a protective cordon about us. Hurrying us along, yet holding to our speed.

The pillar came down again, and this time it transfixed one of our escorts, ramming through the creature's exterior. We watched, horrified, or at least I was horrified, and I heard Ste Etienne curse again. For a moment the impaled creature fought industriously, internal and external limbs clutching for purchase against the spear that had struck through it. Then . . .

I had to play the recording back later, to understand. The impaling rod really was a spiral of separate elements, and they had a hinge-point somewhere above. The looming monster had applied force at the long end of the lever, and the short ends, here within our escort, splayed out with their own abrupt and murderous force. The creature's exterior was prised apart, its struts snapping and plates cracked across, then held open like a dissection subject. A writhing tube slithered down from above, tipped with busy teeth that began tearing into the inner beast within the sundered cage.

We fled, save that the pods could only manage a plod. Ste Etienne was still learning how to pilot one, let alone trying to keep two of them on the move at once. It was stop-start, and the surrounding beasts could plainly have outdistanced us. Yet they would not do so. We had become theirs, and it seemed they weren't going to abandon us, or surrender us to a greater predator. They kept turning and flaring their faces, the displays paired with spikes of static that surely meant they were sending warning signals – some kind of electromagnetic blinding perhaps. But we were very obviously too slow, and the thing above was still hungry. Its feeding apparatus had collapsed and been retracted, and we could track its progress above us by the pieces of briar it dislodged. The invisibility of it, otherwise, was the worst thing. Being unable to see that gleaming beak before it plunged down amongst us.

Without warning, or any obvious signal, half of our escort abruptly left. Not fleeing for their lives, though I wouldn't have blamed them, but turning back. At the fringes of our light, we caught them scaling higher amongst the brambles, moving with that jerky, deliberate gait that didn't seem to care if it was climbing or travelling on the flat. They were heading upwards towards our antagonist, to fight, or just to distract it. They were giving us time.

A lump formed in my throat. I couldn't contextualize it in any way save a selfless act. *Save the visitors from the stars!*

We lost sight of them almost immediately, of course, but the awful looming predator was left behind as well. I tried to imagine them locked in combat, but the physiologies of the pugilists involved were just too alien. My mind defaulted to Earth-standard animals, to recreations of dinosaurs and invented monsters.

Then we were heading downwards, a slope steep enough that we skidded and almost rolled. We lost the connection with Bartokh, but he was right behind us as we jolted and banged down the slope, every drop slamming the breath out of me in the punishing gravity. On the third impact I lost consciousness briefly, and the couch woke me with what I felt was an exasperated air. *Are you sure you're cut out to be an interplanetary explorer?* And I wasn't. I knew that now. I wasn't supposed to be in this pod under any circumstances. That had been the deal I'd made with the universe, as well as Opportunities and Special Projects. I was just someone who made sure meetings happened.

We had come to a stop. Underground. The lamps died, then flared brighter as Ste Etienne did something to them.

Underground. Real ground. Rock, in fact, and mined. A cavern. One camera showed us the round-sectioned tunnel we'd entered by, so regular I'd have said it was artificial. It was artificial, of course, just not carved out by human hands.

Our lamps couldn't reach the far end of the cavern, and no sign of a ceiling showed at the upper fringes of our screens, though we'd already discovered our shortcomings in that respect. There were slanted pillars of stone gleaming here and there, structural support for whatever was above. They shone in our light, veined with metal. Iron, I assumed. Much of Shroud's gravity was due to its big core, not its size, or that had been our best guess. There was a lot of iron that vulcanism could churn gradually into the upper reaches of the world's geology. This appeared in various mineral compounds, and in its pure form, unrusted, because there was no oxygen to corrode it. Iron, and these creatures had just carved straight through it.

They had pulled back from us now, all but one. A sick one, perhaps. There was something different about its appearance, though my eyes couldn't quite process what it was. It wasn't moving with the same animation as the others, just crouching alongside us.

A captive? A slave taken from some other alien polity? No, these were human interpretations.

'Well?' Ste Etienne's voice startled me. I'd lost myself in the images on our screens. 'Director? Bartokh? You there?'

'I am, albeit somewhat battered,' Bartokh's voice came to us. Down here there was less static. The roar of the outside was blocked and we could avoid the channels the local excavators preferred to use.

'Fine,' Ste Etienne said. 'You two put your heads together and work out what's going on. I have about a hundred red lights to look at, across both pods.'

'How are we supposed to do that?' Bartokh demanded peevishly. 'We are literally the first human beings to see any of this. The product of an entirely alien world. I can't just look it up in some kind of manual.' He had been rattled around, and left alone and ignorant for long spaces of time, but I still wished he could be just a little less distant and dismissive. The three of us were all we had.

'Tell you what,' Ste Etienne said, 'I'll play study-the-aliens and you fix the pods. How about that?'

'Well, fine,' came Bartokh's sour reply. 'You are in danger of a demerit when we make contact with the *Garveneer* though.'

'Director,' Ste Etienne said frankly, 'if we ever see the inside of the *Garveneer* again it will be because of me, and I'll be promoted to director of a whole new mission. And you will personally carry me around the inside of the ship on your shoulders you'll be so fucking grateful.'

I actually held my breath. Despite everything, despite the truly extraordinary situation we were in, I was still enough of a slave to my socialization to be horrified.

As though politely letting us air our grievances and settle our differences, our alien hosts had been milling about at the edge of our light, but now a clutch of them stalked forwards. They ended up around the creature which had been left with us, prodding and nudging it until it stumbled into motion. Observing it alongside the others, I spotted the difference.

'Have they . . . shaved it?'

'What?' Ste Etienne asked blankly.

'It doesn't have the hairs . . .' The shimmery aura of reflected light that surrounded the others was absent on this lone specimen. 'The—'

'Antennae,' the engineer filled in. 'Transmitting and receiving antennae. So it's . . . blind? Mute? What?'

'All of the above?' I pivoted a camera to follow, watching as the faltering creature was led out of sight. 'Can we follow it?'

'No idea. Let's see if they let us.' Ste Etienne set both pods into motion again, far smoother on the solid surface, and we kept the beasts in sight all the way up to a gleaming bowl fixed into the rock of the floor. The carved floor, I noticed. The rock surface had been shaped and grooved, a complex design like a spiral of spirals, leading into this bowl. Only here at its conflux was the pattern obvious.

'Um,' said Ste Etienne. She'd stopped moving us forwards, Bartokh's pod clumping to a halt a step later.

'What?'

'I mean, does that look to you what it looks like to me?'

'What?' I demanded again. The monsters were leading their shorn sibling into the bowl. It rippled, and I realized some of the gleam was because it was filled with water, previously still and clear. Or liquid resembling water, anyway. Because it was minus thirty out there, and pure water would have frozen solid.

'I think,' said Ste Etienne tightly, 'we're in for some old-time religion.'

'What are you talking ab—'

They tore it apart. Literally. The escort latched onto the shorn creature, hooked onto the rim of the bowl for purchase, and simultaneously exerted some brutal mechanical leverage. Their victim's casing splayed outwards, exposing a writhing, pallid wormlike thing inside. The actual alien, the pilot of this organic device of levers and rods that served as its outer shell.

More creatures skittered from the darkness, falling on the beast as it twisted there. The details of its dismemberment were mercifully hidden from us, but they devoured it. They must have ripped it to shreds. The liquid went muddy and then milky with its inner fluids, curdling briefly before the creatures drank it all up through pulsing hollow tongues.

I didn't accept Ste Etienne's suggestion, that this was a cannibal sacrifice to unknowable alien gods.

I didn't have an alternative explanation.

The pool refilled. I couldn't see how, but there must have been fine holes in its base, connected to some kind of reservoir. New clear liquid gurgled up as though from nowhere, until the level was near the brim once more.

Ste Etienne got us both moving again. Away from where we had been corralled with that shorn victim. For it seemed we had not been saved from the predator because these monsters recognized us as sapient visitors from another world. Rather they saw us as something unique to offer to the gods.

We were still slow, but they didn't do the obvious thing of fencing us in with their bodies. Perhaps, given our halting progress, they didn't even realize we were trying to escape. For a moment I thought we would be able to just sidle out of there, slipping between the cracks of interspecies relations.

Then they came and took Bartokh. It was as simple as that. One moment his pod and ours were walking in jerky lockstep towards the tunnel we'd entered from. The next, a pack of the monsters had muscled in between us, breaking apart our link. His pod stopped moving instantly, and then began an uncoordinated stagger as Bartokh tried to reassert control. Ste Etienne cursed, started to turn us, then put us in reverse, readjusting her control and which camera view she was operating from.

'Arms,' she told me. 'Get on the arms. Flail, hit them, something!'

I made the pod's mechanical arms lash about, sure enough – no real skill needed for that – but it didn't help. We were already being left behind as Bartokh was hustled off. I caught sight of him, his pod's legs off the ground, cycling uselessly, as they tipped him into the bowl. I started shouting. Screaming at the things, even though the only ears I could reach were mine and Ste Etienne's. Then my reasoning came back to me. I went onto their wavelength and broadcast a long, loud blast of meaningless signal that actually halted them for a second. Gave us two steps of catching up, and Bartokh a moment to scrabble helplessly. But he was half submerged and unable to find purchase against the curved inside of the bowl.

Then they anchored him down, and I watched them strain to open up the pod. The hull held, though. They tore two legs and an arm from it, with the monsters falling back and rolling across the ground when these came away. Their outer skeletons flexed to cushion the impact and they righted themselves easily. Ste Etienne gave a bark of bleak amusement – the triumph of good Earth engineering against the alien menace!

They were still holding him, though, their locking claws hooking onto every irregularity of the pod's exterior. I saw maintenance hatches popped, outer plates buckling. In the back of my head was my empathic recreation of what Bartokh must be going through, alone. I tried to fight it back because the very idea flooded me with misery and panic.

We were closer now. I lashed the end of one remote arm across the back of a monster, and was rewarded by the sight of a delicate manipulator clamp flying into pieces from an impact the creature barely seemed to register.

One of the other creatures had lumped its forward segments atop Bartokh's pod, the myriad of small limbs at its snout end investigating the outer shell. Abruptly one of our screens froze, and then zoomed in as Ste Etienne increased the magnification.

'God,' she choked out.

It had spanner hands. Or it had limbs that terminated with a socket the precise shape of a standard disassembly tool, just like we'd seen in Forty-Eight's recording.

With the same deliberate focus it used for walking, the alien began unscrewing the mountings of the pod's top hatch. The big, obvious hatch that led down to the hollow space where Bartokh was.

Ste Etienne tried to intervene, pushing forwards, just to interpose our pod somehow. But we couldn't even get into the bowl. With almost lazy contempt the nearest creature pushed us away with its hindmost leg, whilst it still latched onto Bartokh's pod with three extended limbs.

I hit it, but the pod's arms hadn't ever been designed as melee weapons. They were for the small and careful movements of research work. I couldn't get any real force into the strike, and when I used a claw to hook into the creature's skeleton, I couldn't even shift it from its anchoring points.

'Brace,' warned Ste Etienne. I didn't, because I didn't understand. Then they finally opened Bartokh's pod up, the entire hatch plate sliding away. Bartokh must have been killed instantly – the shock of Shroud's high-pressure atmosphere at the head of a long queue of fatal conditions. Then came the explosion. Sparked by some sheared wire, some sundered electrical system, plus the oxygen within the pod and the high levels of hydrogen outside.

Ste Etienne had been expecting something huge, I think. Because it was the same mix of clashing atmospheres that had torn open the *Garveneer*. The difference was the pressure gradient here, though. On our ship, when the tanks had ruptured, the compressed atmosphere of Shroud had roared out and carried the explosive force of the blast through our poor, Milquetoast Earth air. Here on the planet's surface, the intrusive, explosive material was at a far lower pressure than the surrounding gas. Shroud flooded in and the thump of flame was almost entirely contained within the pod, as Bartokh's hoarded oxygen incinerated itself into a paradoxical haze of water vapour. This devoured the man's crushed remains, but didn't enact the conflagratory revenge upon his killers that Ste Etienne had perhaps hoped for.

Then it was our turn. The blindly questing snouts of the Shrouded weaved in our direction, barely singed. Ste Etienne put us into reverse again. Not that it would do any good. Not that they couldn't outpace us easily, without even breaking into a sweat, if sweat had even been something they broke into. All they needed to do was interpose their bodies, and clog the tunnel. We were trapped in their terrible chamber, buried beneath the surface of a freezing world the sun never touched.

2.3 LIGHT

We ran.

I say we 'ran', but even on a flat plain we would have lumbered along, leg after independent leg, like a drunkard tripping over their own feet. And Shroud didn't do flat plains. At first we were climbing up through the twisted tunnel in the rock, and I had to help by bracing the pod's arms for purchase, to stop us rolling all the way back down. And at any time one of the creatures could just have . . . stood in our way. Not even pushed back, just stubbornly *been there* and we'd have been completely trapped. Yet stopping us in our foot-dragging, incremental escape didn't seem to occur to them.

Once we made it miraculously out of there, the lamps then showed us the tortuous, criss-crossed three-dimensional landscape ahead, the network of briar-stems sieving the air with their hand-like fans. The whole porous cage of it, an unearthly forest of unknown scope, breathed in unison like a single massive lung. Ste Etienne drove the pod forwards, and sometimes we climbed and sometimes we fell. Always below us was the unknown depth, the abyss that our lamps could make nothing of. Fighting our way through that impossible terrain, Bartokh dead behind us, the whole world's conditions inimical to human survival, I thought I knew what rock bottom was. But then I caught a glimpse in the rearward camera, at the fringe of our lamplight. They hadn't just let us go. The Shrouded were stalking us with a literally inhuman patience. They could have caught us in an instant, I knew. They were adapted to this hellscape, each limb tipped with some kind of clip or hook that clasped the bars of the briars' stem cage and then released at need. We had great stomping feet and feeble spindly arms, and our best human designs had already proved lamentably inadequate to conditions down here.

At last we came to a halt. We had ascended, according to the altimeter – assuming the altimeter was even working properly. If it was, we were around ten metres higher than the point where we'd landed. I wasn't sure why we'd stopped. The lamps picked out no landmarks, save that perhaps one species of briar was giving way to a more robust breed, whose stems sported alternating florets of something pale and fungal-looking. I checked out the radiosphere but there was no magical signal from the *Garveneer*, just more of the appalling broad-band bedlam that was the entire living system of Shroud doing its thing. Amongst this, my algorithms were now picking out some familiar and unwelcome voices.

'Mai,' I said.

Nothing. I craned forwards, feeling every innard strain. Seeing her there, between my feet, head bowed.

'Mai, why have we stopped?' I ran through what I could access of the pod's diagnostics. Plenty of red lights but not in any essential systems. No mechanical failures, and I'd have expected her to be vocal about them if they'd arisen.

'Mai, what's happened?' I asked.

She made a sound. From her it sounded inhuman, as

though some spectral denizen of Shroud had seeped through our hull and possessed her. Not even a loud sound, but something so beyond my understanding that it chilled me.

I saw her shoulders shudder, and she had her face in her hands. Before my eyes the invulnerable shell that was Mai Ste Etienne cracked open and she wept.

I couldn't even reach out to her. I managed to undo most of my straps, but the sheer effort of sitting up left me dizzy and gasping for breath. I collapsed back down, staring up at the close curve of the pod's interior. Staring at the display of lights admonishing me for my vital signs and the undone straps.

'Mai, it's . . . I'm here. It's . . . It'll be . . .' The platitudes. The vapid human gambits we use to console one another. *It's fine. It'll be all right.* It wasn't fine. We wouldn't be all right. There had been a series of traumatic, devastating events that cast us down the catastrophe curve into territory where nothing would be all right again. Half Special Projects were now dead, and if their respective deaths had been quick, they were no less terrible for that. And we were lost, literally and figuratively. Doubtless up on the *Garveneer* they'd already written us off. And why not? We had no way to tell them we were still alive, and what could they even do if they found out we were?

Perhaps, if the commercial exploitation of Shroud continued, then in half a century's time some prospector drone would find the dead shell of our pod. Assuming there was any part of it that the fauna of Shroud couldn't metabolize. This lonely memorial was literally the best-case scenario I could imagine – that somebody might find out what happened to us eventually.

Not something I could say to Ste Etienne to fortify her spirits right now.

'Mai, listen to me,' I said. 'Please. We'll find a way. We—'

'You don't have to,' she growled. I'd left it long enough that she'd pulled herself together. 'Just – can you look at the – the . . .' Some minor system she'd trust me to analyse and repair probably, but her voice broke so abruptly I almost felt the physical shock of it, and she was shaking and sobbing helplessly again, furious with herself. Trying to shut herself up, to restore that iron control she'd always possessed. But the more she forced it, the more it broke her.

So I talked to her. I said all the platitudes, the things we both knew were nonsense. How we'd make it. How we'd raise a glass to Bartokh back on the *Garveneer*. How we'd beat this monstrous world, with its dark and its cold, its crushing pressure and its monsters. All that rubbish, which my mouth just trotted out. I wriggled myself along the ballooning length of the gel couch like a soldier in the mud between the trenches, squirming, shoving myself with my shoulder blades, until I had a foot hanging off the edge. Then Shroud gave me a hand, hauling hungrily at my leg with its devouring gravity so that I slithered another joint's worth over the brink and had to clutch at the gel to avoid going over entirely. After all these painstaking and painful acrobatics, I finally ended up with a magnet-booted foot on the top of Ste Etienne's head.

'Ow!' she barked, jolted out of her misery by the assault. 'What the hell?'

I twisted awkwardly until I had the instep of my boot on her shoulder. With extreme, laborious effort I lumped it up and down, feeling as though the entire limb had been transmuted to lead.

Ste Etienne went still, considering. 'Are you trying,' she got out, 'to pat me on the shoulder?'

'I am *trying*,' I said through gritted teeth, and around strained breaths, 'to show solidarity.'

'It's not working,' she told me.

'I can see that.' I writhed about on the couch some more. 'Can you . . . give me a shove. Back up.'

'You're kidding.'

'I'm sorry. I didn't think this through. Please. The edge of the couch is really biting into the underside of my knee.'

She put her hands around my boot and, with a ferocious oath, shoved me upwards enough that I could get myself fully back on the couch. I slid back into the original dent I'd made, and did the straps up again. I felt horribly embarrassed. I had one job and had just entirely failed to perform it.

Ste Etienne made another noise. Equally unexpected, in this horrible place. She was laughing. It wasn't exactly happy laughter, there was a lot of despair in there too, but it was a healthier catharsis than the crying had been.

'I could have had Skien,' she said. 'Or Rastomaier. I mean, damn me, I'd kill to have Jerennian around now. Except for the amount of room he'd take up. But I've got you.'

I said nothing. It hurt.

After a pause, in which her ragged chuckles were crushed flat by the gravity, she said, 'Sorry.'

'It's all right.'

'It's not. You're doing fine.'

'We're both doing fine,' I said. 'As well as could be humanly expected, honestly. It's just that, planetary conditions being what they are, that's not actually good enough.'

'Lunar conditions.'

'What?' I blinked. Even blinking was an effort.

'Lunar. This being a moon.'

'Oh, right.'

'Important to be accurate.' Ste Etienne giggled again. There was a great deal of 'or else we'll cry' in that laugh. The explosion and deaths up above, the struggle down below, the betrayal by the homicidal locals, all of it had been a raw adrenaline cocktail for us. And now we'd stopped, since we had literally nowhere to go, it had all caught up.

Why wasn't I weeping and wailing and rending my garments, I wondered? It would have been entirely reasonable. Allowed, under the circumstances. Nobody would give me a demerit or decrease my wage-worth for an unproductive display of emotion. It was, I realized, because of Ste Etienne. She needed me to remain steady, so that she could find her own equilibrium. And, in being steady for her, I was fulfilling a purpose. I had something to cling to.

'Right,' said Ste Etienne after a while. 'Enough of that.' I could see from the readouts that she'd had her couch fabricate something potent to shoot her up with. Mood stabilizers were a work tool that everyone in the employ of Concern was intimately familiar with. 'I don't see any of the bastards, anyway. They given up?'

'No,' I said flatly. 'I've kept an ear on the locals' favourite band and they're out there, close behind. Just not so close our lamps can find them. Based on past data, I think they're back in the direction we came, but definitely not just lurking about in their nest. Holding distance, I think. It's hard to tell how much the signal strength is from proximity and what's from their numbers.'

'If they wanted to come fuck us up, they could just do it,' Ste Etienne pointed out.

'I don't know. There's no O_2 in the atmosphere here. What happened when . . . when what *happened* happened – the explosion.' More of a squib than an inferno, but still. 'Maybe it hurt them? And so they're wary of us. Or maybe they just want to make sure we leave their territory or something.' I was well aware that, as before, every guess was steeped in my human Earth-born viewpoint, but what other tools did I have to try and understand?

'Well,' Ste Etienne said, a word without connections. And then, 'Right,' and 'So.' She shook herself, or took her best shot at it under the constrained circumstances. 'I'm going to send you a to-do list of systems I want you to go into the architecture of, the programming architecture, and see if you can iron them out. Stuff where the code isn't meshing with the hardware properly. Can you do that?'

I had no idea, but she didn't want to hear that, so I said I could. It was obscurely flattering that she'd trust me with it, except I was literally the only other human being in the world, and she was reserving the difficult stuff for herself.

'I'm going to use the two arms that are still fully functional to physically repair the other arms, and then one of the legs which is twisting in a way I don't like,' she told me. Fair enough, I wouldn't have known where to start with that.

We worked in silence for a while, and I found I could actually do most of what she wanted me to. I was everybody's understudy, after all. Jerennian would have been ten times better at it, for sure, but I was an infinite number of times more physically present than he was, so I reckon I won out on that point. Outside, the pod's various limbs clattered and banged as Ste Etienne moved them about. I began to appreciate the versatility of its design, for a vehicle that was going into such an unknown. The collar around the pod that had the legs fixed to it, with the main body of the pod designed to pivot freely within it, ensured that we, the occupants, were kept the right way up no matter what. An important consideration given how much work the cushioning couches were doing to support our fragile Earth-evolved bodies. There was another collar, entirely free moving, that bore the various extendable arms, so Ste Etienne could use them to get at any part of the pod's exterior she needed to, as well as any proximate part of Shroud we wanted to sample or molest.

I was about halfway through my task list when I glanced at the screens again. For a short while, the simple focus on doing explicable tasks at a human level had almost driven the threat we were under out of my mind. But the uncompromising light of our lamps was showing some familiar movement. The Shrouded had overcome their reservations and were now drawing closer in that stop-start fashion of theirs. The traffic on their band was picking up too, as though they were psyching themselves towards a new atrocity.

'Mai, you might want to finish that leg soonish.'

She'd been similarly absorbed, but now she looked up and made a hissing noise when she saw the problem. 'Yeah, leg's done. Had a premonition. Did it first. Outstayed our welcome, you reckon?'

God, I envied her cool, even though I'd seen it crack entirely open not so long before. 'Let's just move further from where they actually live,' I suggested, hearing my own voice shake.

She put the pod into motion again, and this time our progress was markedly smoother. Some of the things she'd had me tinkering with had been the algorithms that governed the pod's limbs. A human pilot couldn't be expected to micromanage every footfall, especially if we were going to be actually escaping things. On that basis, there were whole strata of intervening systems that could take her piloting orders and translate them into intelligent movements of the pod's legs in real time, based on a best-fit analysis of the terrain. It was far from perfect, but I liked to think even the Shrouded would notice the decidedly more athletic way we took off right then.

I waited for them to recede into the darkness. They didn't. Instead they matched our increased pace without any difficulty, even gaining a little, almost a whole body-length within our lamplight now. They weren't exactly snapping at our heels, but to human eyes they were very definitely chasing us from where they lived. Which was better than trying to open us up to get at our tasty contents, but I still wished they'd just fall back and leave us alone. The graph of their electromagnetic activity kept spiking, like they were shouting at us to go away.

So we did. While earlier *We ran* had felt appropriate in tone, this time we were just backing carefully away, keeping our rearmost eyes on the enemy. And going where? In my mind there was literally no destination on the entire moon of Shroud to head for. Ste Etienne might have had some revelation, but had yet to share any such oblique thoughts on this with me.

The Shrouded followed us, with the uncomfortable feel of a pack of predators waiting for their prey to tire. Perhaps there were some beasts on Shroud that could employ explosive chemical defences, so the detonation of Bartokh's atmosphere and a little sparking wasn't actually novel to them. Maybe this was how they hunted such creatures, running them into exhaustion. In which case, they'd be disappointed in the short term, because the pod had a fusion reactor and a lot of rather unpalatable supplies. In the long term, though, the logic would win out. We couldn't go on for ever. Nothing could. Things fall apart, as the old saying goes.

They were creeping closer. Playing Grandmother's Footsteps with our heels, never quite nipping at them. We tried different directions and complexities of terrain, still in the hope they were just trying to chase us off, to herd us away from their home. And who would want us, exactly? Humans are terrible house guests, as the vast commercial infrastructure hanging in orbit over Shroud demonstrated.

When they had first appeared sapient, I will confess to an unprofessional pang of guilt at what the *Garveneer* was tooling up to do to their homeworld. But right now, with Bartokh dead, they had it coming to them as far as I was concerned. Equally unprofessional but I plead traumatic shock.

They fell back a little and my heart leapt. We had an understanding again, humans and aliens. *That's right, we're going. You can keep your buried sacrifice temple. We don't want it.*

'What are we approaching?' I asked, because maybe they were actually backing off because there was a sudden drop ahead or something.

'Same old same old,' Ste Etienne said. 'Or . . .' The screens flared briefly as she did something to the lights. What had been mere gradients in the pale greys leapt out with all the vibrancy of washed-out blues. Things that looked like barnacles, or discarded pieces of shell, or . . . alien things. Little glued-together alien trashpiles that could be luxury housing for something a lot smaller than our pursuers. They were markers of some shift in biome, perhaps. The brambly stuff around us was certainly different, thicker and closer woven, the stems scored with spiral grooves.

We'd paused, and the Shrouded began their creeping-up routine again. Their channel was very busy, lots of pulses, in fact a lot of three-pulse patterns as though they were trying to sucker us in again. We didn't reply, needless to say. I reckoned we'd said all we had to say to one another before the murder.

'Onwards?' Ste Etienne checked.

'Onwards,' I agreed, because if the monsters didn't want to go this way then we did.

We made even better headway on the thicker stems, and there was far less shaking and cracking. Ste Etienne even grudgingly said I'd done a good job reconfiguring the leg algorithms. We were feeling good about ourselves right then, despite our lack of long-term prospects. And the monsters were further back now, milling about but not going away.

'Huh,' Ste Etienne said. I switched views and saw that, ahead of us, the whole interwoven substrate funnelled down into darkness to one side. And there was something else. A shimmer in the air, like gnats caught in the light. Little particles weaving back and forth.

'Give it a wide berth,' I decided, and she grunted her agreement, steering our stomping progress away from the hole. If we lost our footing and skidded down it, there was no telling where we could end up.

The air through our side camera was briefly wild with bright rainbow sprays, a fit of dancing glitter, and I realized we'd touched them. Brushed the very tips of them. The thin filaments that extended in a great sensory fan from the depths of that hole, like the taut strands of a spider's web.

It burst out with terrifying speed. Everything we'd seen so far on Shroud had been slow. Slow enough that the deliberate, inexorable nature of its motions had been a part of the horror. Now we learned that Shroud worked to a variety of speeds, because the thing that launched itself out of the hole – its burrow – was enormous and very swift. A plated worm creature five metres thick, each segment bearing a leg, and the legs arranged in a helix around its body, so it could grip every side of its tunnel at once. Its front emerged as a tight fist, gleaming all the colours of mother-of-pearl in our lights. Halfway through its strike it opened into a fan of hooked blades, and then they opened further, lunging out into an extending spiral like predatory orange peel. It clasped around the pod. We heard the jagged hooks and tips of it squeal and shriek across the exterior, muted but still terrifying. Then it was fighting to drag us down, into its lair, where it could disarticulate us with its horrifying multitool head. It would suffer for it, probably, but we wouldn't be around to point and laugh.

Yet we didn't move. Ste Etienne had done something with the pod's feet, thrust them through gaps in the substrate, engaged hooks, something. I saw the whole surface we were clinging to buckle and twist as the thing thrashed, trying to prise us loose. Segments of it fractured away, revealing a honeycomb structure and shrivelled little things that shrank away from the outside air. The tunnel-monster yanked and shook us, and the couches could only absorb so much of the force. We were slung about in our straps like dolls.

Then our screens filled with monsters.

The Shrouded that had been tailing us didn't want to share their dinner, that was what I thought. They were abruptly all over the creature. Being thrown about so much, I didn't catch much of a view of what they did, but they were attacking it. They flicked out those big arms from inside their bodies, latching on, prising and pulling. Working together, because individually they were nothing to the monster. And this ambush predator probably relied on spending its days safe in its tunnel, rather than being strung out like this. It looked armoured, but that didn't dissuade the pack-hunters from having a go.

Without warning, the game suddenly wasn't worth the prize. The tunnel-monster let go, and we were thrown about all over again with the recoil. It vanished into the depths of its lair and I saw several of the pack creatures drawn with it. So relatively, everyone was a winner.

Ste Etienne unlocked the legs and walked us back from the brink. We watched the dark funnel until it was out of the reach of our lamps – which might not even have been out of reach of the ambush worm, given the opacity of the air. Then we stopped. We were now in the very midst of the other monsters, the Bartokh-killing, cannibal sacrifice Shrouded. I suddenly realized that, in the attack, they had moved very quickly indeed. Their careful pacing had become a maniac scramble, not missing a foothold. And they'd already been faster than us, even in low gear.

I was probably whimpering again, and Ste Etienne didn't tell me to stop.

Quiet descended. Quiet for Shroud, anyway, meaning the constant audible rumble of the world still drummed on the outside of the pod like rain, and every frequency of our radio screamed alienness at us.

By the graph I still had running, I saw more spikes on the monsters' communications band. Three pulses. *Knock knock knock.*

I hunched forwards. Ste Etienne twisted round. On our screens, the monsters shuffled and picked their way about, almost aimless save when they all – every single one of them – turned to us to repeat the signal.

She sent three pulses back to them. Like a prisoner knocking on the pipes.

They each replied, simultaneously still for that second, before continuing their restless patrol around us.

2.4 DARKNESS

This is how it was before. When it first came to me, the Stranger. This is how it all begins, for me.

It lies there, trapped, helpless. In this, my first encounter with it, I find it like a stone, like a dead thing. I have been given time to investigate it. No living thing I know was ever built so. Two bulbous segments and a narrow waist. Limbs placed about each segment in a way that speaks of damage or malformation. Half of it is burrowed in, as though it tried to hide itself and then gave up. Making motions with its limbs that achieve nothing, as I gather around it. And then:

It speaks.

Not as I speak, at first. Just the buzz buzz of an animal mind. What had been a dead puzzle like stones becomes a living one with that voice. It speaks in small voices between its pieces. My thoughts touch its outer shell, and return to me scattered a little from all its metal. I build my idea of its full shape by way of the ear. By brush of limb. By spreading my mind so that the Stranger becomes a construct in the midst of my understanding, and I can learn from the gap it makes in the world.

Call.

Response.

Now I have enough perspectives on the Stranger to properly consider what it might be, as I draw my resources in from the surrounding countryside. From close enough, every part of it tells my thoughts that it is radiant with motion, yet all of it sounds still and dead. I have no direct experience of a thing such as this, but I have memories of recent events where ambassadors from Otherlike minds have written to me of them. As though there is a plague of newness come to the world.

It begins to move. I speak to it, but it cannot hear me. My thoughts measure its exterior but it cannot feel me. My expanded perspective lets my thoughts explore the environmental damage around us and I conclude it has come from above. A flier, but unlike anything I have brought down for study. A flier from the highest reaches of the sky, but no more than that. A thing. An animal.

Its limbs whip about, too swift to process, an unwise burning of resources. It ignores its surroundings to instead clutch at itself. I decide that it is injured from its fall. Yet I know fliers. Stick and gossamer things. My thoughts have investigated this stranger and it sounds dense as rock on the outside, the inner hollow of its living tissue buried deep, so I have enough perspective to conclude this does not make sense for a thing from the sky. Once again, I am at the limit of my ability to analyse what I have found here. I need to go home and consider properly.

There is a distinct change to its living-ness, to the vital stillness of its exterior, in a way I have no reference for. I wait for its impossible bulk to float upwards or some other prodigy, eager to glean any secret of its being that might be learnable. But it just lies there, softly calling out in many voices, all unlike anything I know.

I come nearer to it. My thoughts scatter from it in unfamiliar ways that yet suggest a curious logic. As though its dense outerness is designed to evade the attention of a predator I have no name for. Tantalizingly meaningful, as though everything about it would fall into a familiar pattern and become comprehensible, if I could only find the correct configuration of perspectives. I direct some sensory projectors to concentrate their thoughts on it, while others close themselves off from the world to examine how those thoughts come back changed, returning from the Stranger. Around the circle of myself, I build a model of what it is, and it remains incomprehensible. Built wrong, made from the wrong things, speaking a wrong way, alive yet unlike any previously encountered life. From above.

But there is nothing above. This is what I have always understood. Past the fliers there is only uninhabited void, unrewarding of study. One can cast one's thoughts into infinite depth and touch nothing, and no voices sound from that over-abyss. Nothing comes from there. Until now.

My need to return home intensifies, to where I can consider these things. I stand at the edge of concepts too huge for my grasp. Even with all the cognitive resources I have drawn together in this place, I only know there is a huge truth here that I cannot yet comprehend.

It contracts, the bulbous segments drawing together, a defensive stance. Its legs are swift but clumsy. I focus on their movements, refining my understanding of them until I can model their structure. Awkward, ill-suited for the environment they have found themselves in. A thing made for another place. Some inconceivable sky-habitat where such limbs and motions are, perhaps, as natural and sure as my own are here.

I understand that it has contracted its two halves together because it registers my presence at last. It fears predation. And yet it continues to cry out, many thoughts at once in different voices, all nonsense. Some are harmless, some trying to ape a civilized register, others calling in ways that invite peril. My understanding expands and I know this place will not be safe much longer if the Stranger continues to wail. It does not know it is best, if one is hurt, to limit one's cries to the proper registers, so one might be heard by friends.

I try to warn it in words, but there is no reason why this skyborn beast would know such words. It is not even an *Otherlike*, a true mind severed from me. It is just some animal.

I try to chastise it with gestures, reaching out to find some sequence of contacts that might serve as a caution. I threaten an assault, touching its shell with the tools I would use to unshell it, in case fearing me will silence it before it can call ruin down on us.

Then it speaks again. In the register of true mind, as though it is an Otherlike – a thing such as I am, a civilized mind. No words, just *Unh unh unh*, but speech! I am still, silent, dumb, losing all sense of it. It is not merely an animal, a skyborn beast. Or if it is, then there is something in the sky like me.

Unh unh unh! it cries. A tongueless grunt, but in a way that is like true thought.

I try to include it in my consideration. Every part and perspective of me invites the Stranger to partake of my thoughts. I overwhelm them, I think. They cannot rise to my level. And so I must reduce myself, as the being of superior intellect. I grunt back to it, following its lead. *Unh unh unh.*

Almost instantly it replies. *Unh unh unh unh.* Still just gross meaningless ideas, contributing nothing, save that they come from so strange a thing.

I reply in kind. It is a wounded beast whose pain has somehow found a voice, and I can only grunt back, to tell it that it's not alone.

Its meaning intermeshes with mine, those brute impulses textured over with an echo of intellect. A moment, fragmentary and brief, when it becomes real to me. It is acid-tasting to the mind. It feels glancing and sounds leaden. And yet, within that unknown exterior, it is briefly a thing that touches my soul.

With the palette of thought available to me, I recontextualize this encounter. The Stranger continues to leak out unwise cries, inviting peril, and peril cries back from close by. I cast my thoughts out in all directions and they hurry back to me, informing me of the shape and disposition of further things. Differentiating between the mobile and the static, the near and the far. Yes, a threat is answering the bleeding mind of the Stranger. I form a defensive cordon, tracking the peril. I urge the Stranger to govern the register of its thinking but it still cannot understand me.

The intensity of its bleeding mind flares. Out there, the threat stalks closer, wondering who this is that calls to it like its own kind. The lancer, with long strides. I cannot close the holes in the Stranger's mind. I cannot persuade it to govern itself. And I cannot leave it. I am at the limit of my ability to apprehend the world, overwhelmed by a lack of precedent. As I grow more and more agitated, pulled taut between options, it speaks to me again and again, and I speak back.

One!

Yes, one.

One!

Yes, one.

Two!

Yes, two. Very good.

Three!

And so forth. Longer thoughts, all arriving with the clarity of my own, and yet always without meaning. Poor wounded thing. I continue to meet each utterance with its echo, driven to let it know that it has been heard. Eventually it stops, perhaps because it has counted as far as it can, beyond which no numbers exist for it.

It grapples with itself, strange-scented limbs wrestling, before it breaks into two again, and I understand that it does not flee the danger because it does not want to diminish itself. Half of it is trapped, body too inflexible to wriggle free of its cage.

I have been slow to understand this, hindered by the degree of thought that I can bring to bear. As the two parts of the Stranger fall away from one another I take up the fight, tasting its strange, fizzing exterior, lively to my senses without being alive. Physical touch completes the analysis of its situation that my other senses had fed me with, and I free the Stranger from its snare. Its other half attempts to assist but contributes nothing. I let myself expand into a cordon so the two parts can come together again. And I understand now. They are an ambassador of some far place. A message-bearer carrying complex thoughts beyond anything it could itself communicate, and deprived of speech, so it does not simply lose its purpose as it nears me. An ambassador from an

Otherlike mind to mine. From some other self in the skies, to my self on the ground.

It's a message of sufficient complexity that it must come in two parts, beyond one ambassador's capacity to store. Each element is dependent on the other, so neither can be left behind.

Still it just stands there, clasping itself, as the danger grows closer. Now that I know what it is, and its limitations, I cannot afford to be subtle. I begin to push against it, relying on physical force. Contact is unpleasant, as it always is, snuffing out my senses and interfering with the clarity of my thoughts as my pelt flattens against it. The lancer is almost upon us, stalking above, beyond reach of retribution.

I raise a scattering screen of angry thoughts towards the lancer, trying to hide the Stranger with the sheer fury of my cognition. I succeed just enough. The lancer's lunge skewers down, but instead of piercing the strange hide of the ambassador – despite the ambassador's mind leaking out in the lancer's own voice – it thrusts into my midst and withdraws, baffled by the ferocity of my mind. I sing it false locations in its own voice, complicating the world until it cannot know what is real. The ambassador falls sideways, so clumsy in its movements I cannot imagine how it has travelled so far, from whatever distant place it set off.

The lancer repositions itself, groping past my deceptions. This time I succeed only at cost, as its strike takes part of me instead. Tears me apart and finds sustenance in me. A victory for the lancer, a price I will pay. Some understanding has come to the ambassador now and it is trying to go with me, but too slowly. As though its bleeding mind has lost all control of its palsied limbs. The lancer tracks it. I feel the itch of its regard narrowing on this strange new voice in the world.

I come to a decision; I will attack the lancer. Its body is held far beyond my reach, but I will be a distraction. I will abandon myself to fighting it, even as I shepherd the ambassador home.

And so I am fighting the lancer, its beak tearing me asunder. I am ushering the stumbling ambassador away, wishing all the time that it would quieten the babble of its mind.

Just get home. A gap where my thoughts were.

Pushing the burden, knowing only it is precious.

Not even why any more. Such things are left behind, from when I had more thoughts that are now lost to me.

Knowing only: I must keep this thing safe. Bring it to where I can know it better.

Thus I come home.

Here, below, where my careful sculptings of rock and metal can baffle the voices and thoughts of the world, I am able to reach true understanding. I regain that small part of me which I sent out to scout and forage, and understand that it has discovered something genuinely unprecedented. All too often these limited forays into the outside come back full of a newness which, from the greater perspective that home allows, I then realize is something I have seen before. There is precedent, and so I add the experience to my long history, and can only hope some echo of it will travel with my venturesome bands when they depart for the surface once more.

But this is definitely *interesting.*

An ambassador in two volumes. That is what I thought when I was merely foraging. Reconsidering the facts now, assimilating my errant memories and analysing them for error, I concur with my more fragmentary opinion. Some far-off sky power has sent me a message. I cannot know how I am supposed to answer, but perhaps that information will be encoded within the ambassador.

The strange skyborn emissary is not the only visitor from an Otherlike mind. There is already an ambassador standing by from a nearer neighbour – an estranged self with whom I maintain good relations. There have been many messengers recently. Curious anomalies have been encountered. The word is being transcribed into emissaries such as this, and passed from Otherlike to Otherlike to me, here in my home. There is something spreading into the world that I – none of me – can quite understand. Something that promises change.

Change is a loaded thought. A good and a bad thing. It is up to me to make it something of advantage to me. The most eroded recollections of my history still reach back to the time of greatest change, when so much was lost. If a new great change is upon me, I cannot allow it to steal what I am, and what I have become. I must find a way to understand what this sky-ambassador has to say.

The other ambassador, from my neighbour, stands as one mindless, plucked mute so it won't simply lose itself within my greater network of thoughts. It may contain some wisdom regarding this skyborn Otherlike.

I usher it forwards to my lectern, so I may read the message it brings. It is barely a thing now, a single element deprived of senses, and when I open it up, it can scarcely appreciate what has happened to it.

Its message floods out into the waters and curdles into

meaning. A sequence of chemical words painted there by my neighbour. Requests for supplies, warnings of the movements of predators, word passed on from further Otherlike minds. There's reference to the strange new life falling from our skies, but only a brief mention. Nothing heralding these ambassadors.

The skyborn Stranger is wandering again in its clumsy way. I separate them, helpfully guiding the first segment of it to the lectern. No way of knowing whether the order of reading is important, but with my full mind gathered I will be able to reorder the message after it has been released. The second segment is eager, stumbling after its other half, desperate to be read.

At first I cannot access the words within the first skyborn emissary, but information on how these visitors are constructed has reached me through the Otherlike ambassadorial chain and I have already fabricated the appropriate limbs. A test, I consider. To ensure that the message locked within it is only read by the worthy. The second segment jostles some more, almost overly keen to be informative. I would tell it to wait its turn, but the necessary diplomatic segregation means it cannot hear me.

I finally prise the first segment open, eager for its wisdom. Instead there is . . . shock, pain, damage to my own segments, a sudden warning. The closest parts of me foam furiously to deaden the exothermic chemical reaction that has consumed the ambassador's interior, but it is gone almost in the moment of its ignition. A rare and hostile reaction, destructive, inefficient and hard to control. Only a handful of beasts have mastered it.

In its wake is . . . nothing. A shell that tastes of expired chemistry, and some curious-scented compounds that do not constitute a message. Just a small space within the ambassador's thick hide where something lived, and was consumed.

The circle of my understanding twists and rejoins. My comprehension expands to a knowledge so shocking I am immediately preparing my own ambassadors to send to my neighbours. It was not an emissary at all. It was not some novel shell hiding a segment of an Otherlike mind, or a part of me. Nothing from within my long experience.

Life hid within that shell, this much is certain. The ambassador moved and signalled and behaved within the tolerance of a living thing, albeit dysfunctional and ill-adapted to where it found itself. However, as I analyse the residue left behind by its self-devouring, I realize that the world within the shell was not in any way the world without. The pocket of living that the Stranger's hidden body persisted in was unlike any conditions I have encountered, or even hypothesized. More hostile and alien to me than the depths of the sea or the reaches of the sky.

I truly have been visited by something utterly new to this world. In that moment, with all my thoughts engaged within the seclusion of my home, I can even imagine some thin, fiery world where such strange beasts might grow.

The other segment of the non-ambassador retreats. I bombard it with my thoughts, trying to make it part of me, cajoling it for its secrets. I have failed the test, though. It wishes to leave, and some other mind, some Otherlike, will perhaps understand how to unlock its secrets. My chance to do so has gone.

All I can do is give it part of myself, penance for my failure, and hope it may find greater understanding from the next home it draws near to. To ensure that the next time *I* will not simply repeat my error, I paint its exterior with the scents for interest, warning and the unknown. And then I have myself follow it up to the surface, loaded with all I have learned.

The Stranger leaves. I follow, far enough from home that knowledge recedes from me like the tide. Until I know I must follow, and I am wise enough yet to read the message I painted in scent on its shell: *Stranger, of interest, safeguard.* In parting from the majority of me, I have lost the ability to create such language, but remain just about able to read it. I balance on a thin edge of thinking, clinging to the knowledge of why the Stranger is so important. For its strangeness. It is something new that must be preserved until I am in a position to properly examine it again.

The Stranger moves badly but rapidly. If it was more sure of its footing then I would feel the strain of keeping up with its sudden lurches and turns. I send my thoughts ahead of it, bouncing them past the burning beacon of its metal skin, from which its thousand dumb voices leak. It seeks height, choosing always the upwards path, as though it wishes to return to the sky. It is almost out into the open, and the lower reaches of the everstorm, when it stops.

After that, there is a period when the Stranger simply waits there, mewling to itself, and I watch it. I cannot quite understand why, but I have been left with a remnant of memory that it may have reason to fear me. So I give it space.

I wait for it to speak to me, its thoughts to mine. I have a recollection that it can do this. I think it happened just before, though perhaps not to any great result. I cannot recall the details, I left those at home. It preens itself, the thinner of its limbs busy with some kind of grooming rituals. I try extending my mind to it, sending words to it. Encouragement, questioning, offers of assistance. Basic wordless signals. There is only a single segment of it, after all. I cannot expect it to engage in complex thought or communication.

Agreeably, it sets off again, perhaps reassured by my escort. It is moving better now, and I see it has redesigned or repaired its limbs. Just as I might do with my own segments if their exteriors were damaged, or if the challenge before me was one my current tools were unsuited for. Strange, but not so very strange, then. I engage a little more of my energy reserves so I might keep up with the more energetic pace. It moves as though it has a fresh mission from its own home. Perhaps it is going back there, to take the news of its encounter with me to a place where it can properly understand.

My thoughts, rattling ahead of it on the course it has chosen, return to me with the understanding of danger. A tubeworm lair lies in their path, strewn with hairs set for the dumb and brutish to trigger. If I was less, I would not have recognized the peril, but I have enough perspective, and residual memories of past encounters. The Stranger does not, though. They are making directly for the lair at their new speed. I warn them, bombarding them with my voice and thoughts, but they pay me no heed. They are just a segment, beyond thought or worry.

Growing ever more anxious, I close the distance with the Stranger, trying to indicate from my own agitation that there is something to beware of. Then the tubeworm lair is immediately before us and my own instincts take over, drawing me back. I totter on the boundary between informed thought and bestial reaction, because tubeworms are savage and I fear them. The terrain ahead is scattered with its bait-piles – decomposing tidbits of past kills which the worm lays out to draw in the most stupid of scavengers. The Stranger scatters them with its clumping feet, then halts.

At last! I think, and move closer again, trying to make them hear me. They are right at the brink of the tubeworm's senses. Delicate threads that my thoughts and voice cannot even detect, but which I know must be there.

The Stranger lumbers into motion again and runs straight into the trap. Faster than I can process, the tubeworm emerges and latches onto their metal shell with its tongue. The Stranger clings to the weave of the ground, caught but refusing to surrender. Through its alien nature, I read a desperation I have felt before as the tubeworm drags at them. I must make a decision.

The orders painted on its shell require me to preserve it. But in doing so, I will lose much of my ability to gather information from it, to draw conclusions and to report back. I could retreat now, return home and tell the tale of its demise. Or I can fight.

Some balance of mind within me tilts. I *will* fight.

The tubeworm is a thing that lies for long ages without motion. When it does move, it burns the best of its reserves to strike as swiftly as possible. So I must match its pace. I reach within myself and tap the encrypted energy, the inflammatory well of stored fats, and I am quick. My segments blaze, and the swift thrash of the tubeworm slows for me until it is leisurely while I am like the wind.

I attack the tubeworm. I lack either the weapons or the

numbers to truly threaten it, but what I have, I use. I secrete chemicals that eat into its armoured hide. I send segments of myself into its lair to bite at its lower body, where it is most vulnerable. And I use my hooks, clamps and pincers to prise at the edges of its plating.

The tubeworm does not think. All it knows is that its meal has come at a cost it does not wish to pay. So it retreats, letting go of its prize. Some of me is dragged into its lair with it, to be pulled apart and feasted on. I feel each segment unshelled as it tries to scrabble out.

I am left alongside the Stranger, with words on its shell I can no longer read. But my instinct says: *protect*. I don't know why.

My instinct preserves one tool too. A thing linked to the Stranger. Simple enough that I can perform it.

Unh unh unh. Husks of old thoughts prompt me to this base grunting. I don't know why.

Unh unh unh unh, the Stranger says, and I reply in kind.

We let our thoughts process one another, cut off in the midst of the world.

INTERLUDE TWO

Let me put this into words you can understand. Life in the open air was hard compared to life in the ocean. For an age, things lived a littoral existence on the edge of land and sea. But it was easier on this world than on yours. Here, the line between land and sea is seldom as sharp as it is in your home. On this world the atmosphere is denser, so there is a steadier gradation from the depths of the sea to the heights of the air. A ladder life could climb.

And it was worth it for here the air was wealth.

The countless aeons of oceanic life and vulcanism had enriched the atmosphere with a thick stew of energetic chemistry. Methane, ammonia and raw hydrogen, the stuff of life. And there was more: in the far upper reaches of the atmosphere the energies of the planet and distant sun worked upon the methane that rose up to meet it, cleaving it into acetylene and ethylene. The nuclear fire of this was then banked, so its energy could be released later, once the products of the reaction had sunk back down. The life which left the sea found this wealth, and evolved catalysts to better exploit it. The slow and sluggish ways of the sea were left behind. Crude walking shells made from excess metals were refined into a myriad of stilt-legged walkers, to carry these ancestral explorers across the terrain.

Meanwhile the eternally chasing winds of the Everstorm were discovered and used to distribute microscopic eggs and young. Sessile life spread out across the surface of the sea, building islands, promontories and whole continents of wrack that rose and fell with the tides. Then they built upon those again, flexible and compressible, so even the greatest swell of the sea would be no more than a sway on the surface. And so the rich air could be channelled, funnelled and harvested from them. The first life had a free hand to gather all this wealth, just sitting immobile wherever the richest currents of air were. Energy was encapsulated in highly oxidized compounds, to be tapped with reactive hydrogen and coppery catalysts as necessary. Unlike the history of your home, we never let dangerous, rampant oxygen loose into the world to wreak havoc. Better to keep it locked up in fats and tissues where it can do less damage.

The biomes of the land fell to life, and in so falling, created many more biomes. Every step evolution took split off more paths. Life in the open air swiftly became more varied and complex than the slow undersea world could ever have dreamt of.

But it was still dark. Not the only, or even the most important, difference to the worlds you know. The same richness which made the atmosphere such a feast for life also ensured that not even the most meagre sunbeam could ever shine through upon that host of evolved forms. Without the least receptor to receive light, the cost of bioluminescence was not worth the candle, so to speak. Nothing glowed or flickered in the abyss of this world. Nothing saw. In the sea, where life had even bothered to evolve mobility, it had groped its way across the sea floor, in those regions near the vents where life was worth living. The most active and acute of sea beasts had developed organs to taste the water for signs of what might lurk up-current. In so lethargic a world, expensive sensory apparatus is overrated.

In the air, with life becoming swifter all the time, duelling evolutionary lineages raced each other to apprehend the world around them. Whilst the sessile forms could simply sit and sieve the atmosphere, or cast themselves into the Everstorm and hang in the air cycle with mouths agape, everything else had a keen need to understand what was around it, whether it be food or threat or simply obstruction.

And so the first great innovation: life learned to hear. The sluggish water things had barely needed to, but the dense air conducted sound well. With a few fine hairs to detect tremors, the world became an encyclopaedia of information. The scrape and rattle of a predator's carapace carried to the sensitive fans of its prey. An arms race of stealth and auditory acuity accelerated the pace of evolution.

Simultaneously, life grew. At first, facing the challenge of a less supportive medium, the air-eaters had been small. However, the articulated carapaces that had evolved in the sea worked with the dense air to allow rapid expansion, and competition between predator and prey drove this further. Every time a limit of size seemed to be reached, some new technique of exoskeletal architecture surpassed it. Larger frames and forms then permitted more complex sensory apparatus.

Yet, for another long age, it was still all just groping in the dark, scenting the air, furtive listening. Until the next great leap forward for life. When one creature learned to listen so *hard* that it could hear its own echo.

PART THREE
SEA / AIR

3.1 LIGHT

'They're not killing us,' I noted.

We'd sent another few signals on the Shrouded's band, receiving the same parrot response. The creatures followed our lead in the rhythm and number of pulses we sent them, but didn't seem able to anticipate patterns.

'They haven't killed us *yet*,' Ste Etienne said.

'Cheery. They saved us, though.'

'They killed Bartokh. So saved us for *what* exactly?' She was trying to make herself sound grim, but I reckoned a lot of what I was hearing was exhaustion.

'Do you need me to drive?' I asked.

This was an unexpected-enough offer that she actually levered herself up and tried to look round at me. She gave that up quickly enough, though, collapsing with a, 'Damn me . . . Need you to *what*?'

'Drive. The pod. You take some sedatives and recharge.'

Ste Etienne chuckled and then the lights went off for a second, before slowly waxing back to full. There followed a few subjectively terrifying but objectively hilarious minutes of us both checking our diagnostics until we tracked down the glitch and patched it. Just an unanticipated side-effect of our last batch of code improvements. Not the pod systems failing. Yet.

'Juna, I'm going to say something now, right.' Ste Etienne was lying there, staring up at the concave interior of the pod. I saw from my readouts that she'd just had her couch shoot her with a new wave of stimulants, so no driving for me, apparently.

'Go on,' I said, feeling a little irritated, because if she wasn't sleeping then maybe I could have.

'The beasties don't need to kill us. They only have to wait.'

Yes, I knew this. Yes, it was obvious to anyone, given a moment's objective thought. But it's amazing what the human mind can achieve, in the realms of dismissing the obvious. Right then I found I absolutely didn't want to get into this. Why couldn't we just tromp across the surface of Shroud as though there was a point in us doing that. As long as we were moving there was hope, right? Except she was correct, and there wasn't. The Shrouded might be tame as puppies right now, but the moon never would be. The conditions out there would kill us instantly, and while the pod would keep us alive for a good while, that wasn't for ever.

'We need to find a way to signal the *Garveneer*. They don't even know we're alive.' I left out whether or not they'd care. If we added actual human failings to the mix then the problem would be extra-plus insurmountable, rather than just regular old impossible.

Ste Etienne grunted. In this eloquent wordlessness I read the question, *And how are we supposed to do that exactly?*

'I know we told Bartok it was impossible before, but there must be a way. If we cannibalize every pod system that can contribute, we could maybe reconfigure the arms as a sort of super-antennae and blast out a strong enough signal to make it through the ambient . . . We know it becomes weaker at higher altitudes.' Because there was less life there, compared to what surrounded us and created a constant chorus to drown out our best efforts.

Something bipped, and I saw Ste Etienne had just shunted over some data to me. It proved to be calculations. Very involved ones. I wasn't even sure when she'd had the time to make them, or whether she'd had a modelling algorithm do them in the background, starting the moment she realized where we were. It was a mathematical exploration of precisely what sort of signal we could generate, giving it absolutely everything that wouldn't impact too much on life support, and based on the lowest ebb of the surrounding radiospheric chaos. It didn't even reach half a kilometre, and we'd burn out key transmitting systems after around forty seconds.

'I mean,' she said, after I'd been absorbing this in silence for a while, 'I'm sorry. I should say I'm sorry.'

'I . . . don't think this is your fault. It's the planet.'

'We knew enough,' she told me. 'I should have . . . I don't even know if there *is* a way I could have given the pod enough *pow* to reach orbit with a signal, on top of all the bloody list of other things everyone wanted. But I should have. I just . . . didn't take this place seriously enough. I'm sorry.'

Her voice was shaking by the end, and I wanted desperately to reach out to her. I only had my own voice, though, and my own tenuous emotional stability, which felt as if it was on the verge of breaking. But I needed to hold together, for her, so when I broke later, she could be there for me. We'd take it in turns to fall apart.

'You've done wonders,' I told her. 'Look at all the ridiculous shit we've already survived. We didn't really know anything about what it was like down here. I mean . . . these things!' I gave a tiny gesture with my fingers at the screens, which still thronged with companionable monsters. A gesture she couldn't even see, but hopefully my tone implied it. 'And this briar stuff. I mean, the way we've been able to reconfigure the pod on the fly. It's . . . it's exceeded expectations.' The old phrase that Opportunities used, when they were really pleased with you. 'We'll think of something. This isn't it. We're not done.' The halcyon song of the human spirit. Where there's a will there's a way. I said those words and wondered just how many people in the past – lost in blizzards, in caves, trapped in dead spaceships – had said exactly that. And died.

When she said nothing, I added, 'Come on, that was my top-of-the-range pep talk. You'd not get that from Big Mike Jerennian.'

She let out a bark of laughter, big enough to make her hiss with pain, because sudden movements that the couch wasn't ready for could wrench your muscles.

'Look,' she said, 'I appreciate the effort. Very good pep talk. Exceeded expectations. Except it doesn't change the fact that we have no . . .'

She stopped. And she didn't actually send any data over to me this time, but somehow the idea communicated itself to me.

'I thought you were about to say, before . . . that there was somewhere we could head for. Something . . . ? Or was that just trying to give me hope?'

'Like I'd ever,' she scoffed. 'I mean, yeah. Only all that shit happened and it went out of my head, and . . .' Bartokh's death had hit her far harder than I'd have expected, perhaps because she saw it as something to be laid at her professional doorstep. But now she was *back*, and maybe it really had been thanks to my pep. Ste Etienne had her engineer hat on again. 'I was thinking there is basically one landmark on this entire moon, from the human perspective, and that's the damn elevator anchor we installed. So if they're going to carry on with the exploitation of Shroud, that's where we'll run into human activity.' Meaning mechanical activity, of course, because I reckoned the piloted-vehicle program had been set back a ways, for obvious reasons. 'If they're not actually exploiting, then it's still . . . full of shielded fibre-optics we could use to get a signal straight up out of the gravity well,' she concluded. 'If they're even listening, and if they'd then spend the resources to pull us out. And if they won't, we'll refit the pod to climb the damn cable and the hell with them.'

'That's great,' I said, enthused. 'Let's set course for the anchor and off we go.'

She was silent for a while. I thought she was plotting a course at first, but my readouts didn't show any activity. Eventually she admitted, 'I don't know where it is.'

'But it's . . .' I frowned. 'I mean, it's the pole, right? Or not the pole, but the . . . the outward "pole" pole.' Making little gestures she couldn't see. 'The . . . facing away from the planet bit.'

'Very goddamn technical notation there,' she spat, though there was a bit of a bleak laugh in it. 'Pole doesn't mean what we'd use it to mean on Earth, though. It's not the magnetic pole. And I'm not even sure I can work out where *that* is, given the goddamn row the neighbours are putting out on every EM frequency. But even if I could, the anchor isn't there. And then I wouldn't be able to work out where it is *from* there.'

'The anchor *is* at the point furthest away from the planet though, right? Literally diametrically facing away from Prospector413b?'

She grunted once more.

I glanced at the screens again. The Shrouded were busy doing some task. In fact, somewhere outside the reach of our lights, they'd been murdering things. I knew this because some of them had brought bits of the thing they'd murdered back to where we could see them. Nothing visceral, to a human eye, or really identifiably living. Except they were apparently having lunch, extending flexible proboscises into something chunky and unshelled. And . . .

'Um,' I said. 'They've . . . given us a present.'

Ste Etienne checked the cameras. A lump of something had been dropped next to us, along with a scattering of rods, thatch and interwoven strands which had probably been the outsides of something living not that long ago.

'Or they made us a packed lunch,' I added.

'I'm sending it back with a complaint,' Ste Etienne decided. 'Look, maybe we could . . . if we did move towards the magnetic pole, and I can work that out, it would tell us which hemisphere of the moon we're on, and then we could . . .'

'I have an idea,' I said.

'Yeah?' The degree of doubt in her voice was not flattering.

'I want you to get some sleep.'

'That's your idea?'

'So if my idea works, you're good and fresh to drive us where we need to go.' There was a whole other chasm we were tip-toeing around, because she'd just used the word *hemisphere*. Shroud was bigger than Earth, and the pod wasn't breaking any land-speed records, even on the flat, but neither of us was going to even *mention* that whole extra level of obstacle. Not when there were plenty more insuperable difficulties far closer to home.

Eventually I talked her into taking an anti-stim shot to counter all the buzz she'd had, and then to take the actual sedatives. When she'd finally drifted off, the pod was very quiet. Well, not really. Aside from the faint strained wheeze of Ste Etienne's breath, there was the susurrus of Shroud's audible chorus. This was in itself just a faint echo of the electromagnetic orchestra constantly braying on all channels. Except there were variations. The pod systems had been tracking them, as a side effect of searching for a window through which to get a transmission out. There were no windows, but I reckoned I'd worked out the variations. We had been very lucky with the timing of our crash. And, given all the other circumstances, perhaps we were due a little luck. I set up an algorithm to signal the Shrouded sporadically, just to let them know we were still alive in here, and then I set to work on my own calculations. Being everyone's understudy, as well as the team's administrator, meant I could do everything *a bit*. And I'd been taking the lead on the analysis of Shroud's radiosphere before the accident, since nobody else had wanted it.

So it was that when Ste Etienne's couch decided she was rested enough and jolted her into wakefulness, I said, 'I've done it. I've found the anchor.'

It hadn't been an engineering challenge, in the end. Nothing that called for Jerennian's advanced data manipulation skills, or for a top-flight scientist, or whatever Bartokh would have brought to the equation. Instead it had just needed someone to work with the software tools and do a lot of tedious data analysis. My stock in trade, really. The dull grunt work that had always devolved to me.

The pod, being primarily a human research tool, had been patiently recording a ton of data ever since we got its systems up and running. Business as usual, as far as it was concerned, with its systems being blissfully unaware of our problems. I'd started out by trying to map the magnetic field of the moon, because what else did we have to go on? We had an idea of the geopositional relationship between the actual magnetic pole and what we'd been calling 'the pole'. That imaginary line running through the heart of Shroud, which was pointed towards and away from its mother planet. Of course, even if we were sitting on the magnetic pole, that wouldn't have helped us find the anchor. From the north pole, everywhere is south, and between the EM blizzard and the occluded sky, there was no way of working out which 'south' we'd need.

Except . . . the more I looked at the data, the more I saw something interesting. I could filter out the signals of our murder-friends by then, having had enough exposure to their particular flavour of chatter. This cleared the field a lot, given how much contact we'd had with them. I was left with the rest of the . . . what, bio-radiosphere? The big mess that was Shroud's myriad competing electromagnetic voices, anyway. All the signals and navigational attempts of its inhabitants, clashing and overlapping. But through the random noise of it all, I could see a pattern. A steady rising of background electromagnetic activity that didn't seem to relate to the life around us. It was only Shroud's essential non-Earthness that obscured the revelation for so long.

I realized (I explained to Ste Etienne) that I was seeing dawn.

'What dawn?' she demanded. 'There isn't a dawn. Or didn't you look outside recently?'

'The sun. Shroud orbits Prospector413 – look, I'm just going to call it Prospector from now on. Shroud doesn't rotate, but its orbit around the gas giant means the system's sun crosses the sky of Shroud. Or it would do if the air wasn't so thick. Light doesn't get through, but some EM frequencies do. I can actually pinpoint the system's star in the sky, now I've sieved out all the noise. And the planet, Prospector, also registers. It sends out a crapton of EM activity all of its own, loud enough that you can just about plot the footprint of it in the sky. So by cross-referencing precisely where its EM print is, and the elevation of where the sun goes up and down, and correlating all of that, I can work out roughly where we are. Which means I can work out roughly where the anchor is in relation to us.'

'Let me see the data.'

I showed it to her. She parsed it for a few minutes and then grunted.

'Not seeing much of your damn "dawn",' she noted. 'Sun just peeping at the horizon and then sodding off.'

I agreed this was the case. I'd worked out why that was a problem, but was still trying to put a good face on it all.

'So we're on the side of Shroud that faces inwards, towards the planet,' Ste Etienne clarified.

'Right.'

'Well, shit.'

'I know.' We were on completely the wrong side of Shroud. I could stick a wobbly pin in the planetary map and it would give us a heading, within a tolerable margin of error. I'd then be able to refine our course as we travelled by incorporating the extra data which time would accumulate. But none of this got around the fact we were a terribly long way from the single solitary landmark on Shroud which promised any kind of salvation.

'I don't know about reserves, how long we can last, obviously,' I said soberly.

'It's more to do with mechanical stress, wear and tear,' Ste Etienne said. 'I designed these things to operate under what we knew of this world's conditions, sure. Not to actually tromp most of the way around the moon's circumference, though.'

'Right,' I said again. I'd been working for hours to achieve my little victory, and now my point of view just pulled out further and further to emphasize how very small it had been.

'I guess we'd better get going then,' Ste Etienne said.

'What?'

'Get our feet moving before some damn alien vine grows over us.' She began to bring dormant systems back online.

'But you said—'

She twisted on her couch, trying to look at me, but probably only snagging a brief glimpse of my feet at best. 'I don't see we've got another option, Juna. If you're holding out on me, now's the time to say. Otherwise, let's start walking.'

Our escort fell into step around and behind us as we set the legs in motion again. I was already trying to convince myself I wasn't feeling some minute, grinding irregularities to our motion. Whatever conditions Ste Etienne had tried to take into account with her designs, I sensed the miasmic, organic grit of what passed for air on Shroud would be working its way in to clog every crack and seam of our vehicle.

We'd noted the change in the surrounding substrate before, and we were definitely crossing into some new environment or biome. Soon the weave beneath us started thinning out, enough that we had to reprogram the walking algorithm to avoid the holes. We saw a lot more smaller life too, just fleeting glimmers in our lamps. Things scuttling out from under our tread, or leaping in spring-loaded bursts between briars. Their trajectories were weirdly flat in the high gravity, their descents slowed by unfurling parachutes like webby gills. Everything seemed to have its little constructed shell around it, from the heavy ironmongery of our escorts to gossamer-delicate things like spun sugar. It all moved very slowly, until it moved very quickly, and it was usually only in the transition between those states that we became aware of it.

Then we ran out of land. It appeared the reason the surrounding environment had been thinning out was because we were approaching a patch of open sea. Or perhaps the nearest this part of Shroud ever got to it.

We were on the warmer hemisphere of the planet, after all. Which meant it was still far too cold for water to be a liquid, by Earth standards. But we knew by then there were enough impurities in Shroud's seas, ammonia especially, to keep them from freezing at the balmy minus thirty degrees we were currently enjoying. High summer for Shroud. Except it was not *quite* enough to keep the seas from freezing, whilst being not *quite* enough to freeze them. It was a biome unknown to Earth, and we edged forwards, exploring it with the utmost reach of our lamps.

Around us, the briars arced in looping tangles, taking them out of range of the light. Elsewhere, scouting along the 'shore', we could see where other such vegetal expeditions had come down onto the sea surface and created little pontoons; artificial living islands colonizing the water surface. We picked out a great snarl of bubbles or blisters puffing out the stems where they reached the water. At first I thought these were eggs or parasites, and maybe some were. Ste Etienne suggested ballast, though. After all, there was hydrogen for the taking in the atmosphere. You could get some powerful flotation going by encysting it in bladders.

This prompted us to look back through the rear cameras. We'd felt the rise and fall of the water, after all – surprisingly gentle tides, because of the moon's negligible rotation compared to its planet. The place where Bartokh had died had been mined into the rock, for sure, but perhaps everything else had just been this living weave over the water surface. And now we were far enough away from the rock – that island, I supposed – the living substate couldn't keep up a full coverage.

Where the briars hadn't colonized there was ice, but it was a weird sort of ice. Not a solid sheet, like there used to be over Earth's poles, but a broken-up and constantly shifting skin of it. Jagged pieces chewed against one another over and over, as though Shroud's unpleasantnesses hadn't been complete without a naturally occurring meat grinder. And it was alive, or at least things were living in it. Every time the motion of the ice made a gap, instead of open water being exposed, something living extruded into the air in a blooming fan, then retracted again. It was happening regularly enough that the things below must have had either foreknowledge or control over the clench and crush of the ice, or we'd have seen severed alien gills all over the place. In our lamplight they were a lurid orange, some side-effect of their chemistry, because no native could ever have appreciated it.

'Do we . . . float?' I asked Ste Etienne. I didn't like our prospects, honestly. With that grinding ice, floating didn't seem to be the right question.

'We do not,' she said. 'And I don't like the look of the alkalinity in the air, let alone the water. We have a lot of components that'll start to corrode at that pH level.'

Our instruments were reporting there was a dense layer of water-saturation hazing over the ice, liquid droplets and frozen crystals constantly trading states. It was cutting down the range of our lamps even further. We sat together, alone, and considered the ceaseless teeth of the sea.

3.2 LIGHT

'What do we do?' I asked, after we'd both sat silent for quite long enough.

'Maybe we can use the briars as bridges?' Ste Etienne suggested weakly. I felt the pod wasn't arboreal enough for that kind of climbing, although better that than swim.

'Maybe we can go . . . around,' I said, the words dying even as I uttered them.

Ste Etienne's pointed silence made it plain what she thought of going around the *sea*, on a planet where most of the surface was underwater. And with that water under a varying, unknown thickness and quality of ice.

Our escort, the Shrouded, had gathered behind us, giving themselves their air of a gaggle of children. Now they were moving once more, brushing past us on either side. Or not quite brushing – the shimmering field of metallic hairs around them never quite touched us. They clearly had some manner of electric field sense, to supplement the merely tactile, when it came to knowing their immediate environs.

They were doing what they'd done when we'd stopped before. Foraging. They moved together, or at most split into two or three groups while they were within sight of us. It gave me the impression they thought they'd learned our human ways now. We had stopped, which meant it was feeding time for the Shrouded.

'Huh,' said Ste Etienne, watching them march right out onto the ice. The whole shattered crazy paving of it rippled and bowed under their weight. Each of them must have matched a fair proportion of the pod's mass, given their size and the fact their exoskeletons looked like they had a lot of metal in them. I'd have expected them to go straight through. After all, we could see actual water when the ice parted, and it wasn't that thick.

Ste Etienne made another sound, part impressed and part horrified. Not a noise I'd really heard before and not a welcome one. I couldn't see it at first. Not even after she'd tried to describe to me what was there. In the end she had to annotate the camera view, drawing crude, wobbly lines on the screen until the reality of what it was suddenly sprang to my eyes.

The creature, or creatures, we'd seen sticking their fans up through the ice weren't just isolated marine fauna. What we'd thought of as water appearing between the plates and chunks of grinding ice was actually *all* creature. It hung right below the ice, at the very water margin, a living meniscus. It was translucent and bulbous, a mass of puffy tentacles or vines, glassily clear so the dark water below just showed all the way through it. It must have been buoyant of its own nature, but it was also clinging to the ice using a welter of suckers that extended as far as the light did. The fans which kept splaying out into the air were a part of it, not separate entities at all. It was holding the ice together. Perhaps it was even regulating how the ice formed, messing with the purity of the water to manipulate the freezing point. Subjectively, it was perhaps the least appealing example of Shroud's bountiful natural history we had yet encountered. Objectively, it was our road forward.

'Okay, so I'm going to try it,' Ste Etienne said. She very carefully walked us to where the ice began and put one leg out, reaching as far as possible. There was a marginal zone where the tentacular mass of living material was just a fringe, and we didn't fancy toppling into it, then maybe falling through. I think she had planned a cautious and measured experimentation of weight and balance, but what actually happened was the briary at the edge – the stuff we were unthinkingly standing on – turned out to be thin as well, and we abruptly felt it giving way beneath us. We'd forgotten, in our new discovery, that this wasn't on dry land either. Ste Etienne had a moment of decision and chose advance over retreat, using our back two legs to lunge us forward. Just like 'run' earlier, 'lunge' overstates the case, but we ended up on the ice, which fractured madly beneath us with every step. And yet it refused to come apart, the mass of the creature below absorbing even the pod's compact weight.

We waited for the tentacles, the waking of the kraken. The thing below us remained torpid and static, however, just groping through the air with its innumerable fans. Each of these must have been two metres long and four across at full extension, formed into a fantastically complex mesh.

The Shrouded started to come back, doubtless called by some electromagnetic whistle. They spread further around us than they had been before. An understanding of the load-bearing qualities of the ice, perhaps. When we stopped again, just to absorb the enormity of what we were now doing, they waited. Easy to anthropomorphize them as quizzical dogs awaiting their master. Except they'd murdered Bartokh. Except dogs – ants, mushrooms even – were our close cousins compared to anything that had evolved on Shroud.

But they were all we had, and we could only try to understand them, as perhaps they were trying to understand us.

Life on the ice was not as secure as it had first seemed.

The ice surface itself was gritty with bits that crunched under the pod's feet, and this stuff was alive. Or at least it was before multiple tons of Earth tech crunched down onto it. The ice groaned and settled, and water sluiced over it, so we were ankle deep a lot of the time, but the thing below continued to hold us up. Its substance must have been saturated with hydrogen bladders, but apparently this was a fair price to pay for a monopoly of the water boundary in this biome. Possibly the whole thing, extending over tens of kilometres, was a single entity. The largest individual lifeform humanity had ever encountered. You lose your sense of wonder after you've slogged over it for hours, though.

Here and there a tall, rigid stalk stood out from the water, swaying with the motion of the ice, like the masts of boats, and extending far past the range of our light. They were constructed of interwoven rods which were rooted deep in the creature below us. I did some scratch calculations and told Ste Etienne the small movements we were seeing indicated enormous swing far higher up at the spire's terminus, which must be in the storm wind zone. They might have been some parasite using the ice-dweller as an anchor, but I thought about some feeding mechanism, or maybe seeds. The constant wind-cell cycles could be a reliable way of spreading your young. Ste Etienne, for her part, suggested that none of this actually *helped* us. But then she was doing the driving. I had to keep my mind busy somehow.

We saw occasional islands. The sea here must have been relatively shallow, and maybe the Thing Under The Ice was anchored to the bottom. Or maybe there was a whole aquatic ecosystem hidden beneath it, dependent on its leavings. Wherever land thrust through the ice, creating an enclave beyond the ice-dweller's mastery, there was another spread of the briar stuff, usually linked to cousin colonies by arching bridges leading off into the gloom. The two classes of living substrate didn't seem to play well together, though. I wondered why the briar stuff didn't just extend over the ice. We found out why when we tried to stop and rest.

We should have been working in shifts already, to make the best time. The idea of living a fugitive existence, where each one of us was without human company, had been too much, at first. We never quite discussed it, but we both must have known we were trading off short-term efficiency for long-term mental health, and been happy with that. This meant we had to stop and sleep, when the pod's telltales warned us that the colossal doses of stimulants it had been feeding us were crossing over into the danger zone.

We parked up, but our escort didn't like that. The Shrouded kept prodding and poking at us, reaching out the long, heavy-duty arms they kept folded up inside their hollow armour, and rapping our hull. It wasn't something they'd done before, but we could hardly expect to have their entire range of behaviour nailed down after so short and tumultuous an acquaintance.

We only understood their concern when the ice started giving way. There was a colossal crack and our world skewed thirty degrees to the right, before the gyros compensated. The pod shot us full of stims up to our eyeballs, more awake than we'd ever been in our lives. Our carriage was plainly going down, and we had a desperately hard scrabble at the thinning edges of the ice and the pulpy, bulbous stuff of the ice-dweller. We'd have vanished into the abyss, to sink without trace, no question of it, save for our escort. They had a full understanding of the problem and latched onto us with their big arms, braced against the ice. Then they hauled us free, just as they'd hauled Bartokh's pod out of the briars. Before leading us all to their sacrificial temple.

I checked the rear cameras. Where we'd been settled was a pool of open water. The ice was now gone and the creature beneath had withdrawn. It didn't like anything blotting it out, apparently.

We continued, apprehensively, on the now less reliable ice. Later we did see some ice-adapted Shroud life, and it was either impressive or ridiculous, depending on your viewpoint. We were below the main wind cycle here – calculations suggested there must be a significant ridge of something solid ahead of us – but there were still gusting lateral winds across the relatively open terrain the ice represented. And this had plainly been a reliable environmental factor for a long time. Reliable enough for the new locals we encountered to have evolved sails.

They were long-limbed, spindly things – perhaps struts was the right word, rather than limbs. They dwelled within a woven exoskeleton, like our escort, but their 'feet' were skids with curled ends, just like a sled's. And their alternating limbs managed a set of membranous fans, which they opened and closed to take advantage of the shifting gusts. We saw them flash past us with the wind, little more than blurred ghosts, even when we replayed the recordings. Later we saw them tacking close by and ahead of us, in and out of our field of vision as we forged onwards. Zigzagging up the ice against the weather. Did that make them intelligent? No more than a spider with its web, I supposed. Our escort chased them occasionally and brought one down by some complex encirclement we couldn't appreciate, as it took place mostly out in the darkness.

There were things in the air, too. From some islands we saw stalks or tethers extending off into the dark, which matched images from the drone footage we'd been examining before the accident. Once something buzzed us, or rather the wind brought it incidentally close enough for us to glimpse it. It was gliding surprisingly slowly and our lamps caught the lower edge of it. A translucent wing strung between curved fingers is what it looked like. Very large, so its full extent went beyond our vision in both directions. Rather than being cupped like a sail, so the wind would fill it and carry it along more swiftly, it arched the other way. This actively slowed it to the point it must constantly have to adjust its trim to stop itself dropping from the air. Neither Ste Etienne nor I could advance any explanation for this, and by this point she was simply answering each of my hypotheticals with 'Nothing about this world makes sense.'

We had resigned ourselves to shifts by then. The chemical cocktail bar we had access to prescribed that I sleep for six hours, then we overlap shifts for six, then Ste Etienne slept, and another overlap would follow that. Enough to keep us both rested and sane. Which meant Ste Etienne had to go half-mad making sure I knew how to drive the pod. A process that stripped from me the feeling-good-about-myself I'd accumulated from the navigation trick. But in the end it was mostly just algorithm-wrangling, and the pod more or less drove itself. Until it hit trouble.

I'd slept, and then she'd slept, and woken, and so we were both bright and ready in our overlap shift when the next crisis arrived. Meaning we ran into more of Shroud's locals.

We'd had the sense we were close to land for a while. It was mostly the way the swell of the water had changed. There was a definite back-ripple effect going on, according to Ste Etienne. She reckoned we were going parallel to a shoreline we couldn't see, and maybe we'd make better time on solid ground. Despite our continued progress, I don't think either of us trusted the ice-dweller any more. Around that time, we went through the ice the first time, and had to wrestle with the pod's legs, tilting and splashing as everything below us promised to just withdraw its support and consign us to the depths. By the time we had ourselves back on . . . well, not terra firma, but the closest we were going to get, it was clear we had actually reached the extent of the ice-dweller. There was plenty of ice still, but it was just a mishmash of jostling pieces, as well as a huge berg, ploughing a slow course through the slush of the rest. Probably some calved-out piece of glacier from somewhere far off our rudimentary maps.

We altered course, keeping the trailing edge of the creature below us, trying to find a way around. Which was when we saw the sharks.

Not sharks, obviously, but seeing their steep, dark backs cutting through the maze of fragmented ice spoke to something primal in the sea-fearing human soul. They were coming on very quickly, and the bodies beneath those slicing fins were surely as big as the pod, or bigger.

Our escort got in our way, then. Despite the danger of being stationary, they closed ranks around us. At the time I assumed they were defending us, but what they were actually doing was preventing even the lumbering attempt at escape we could have made.

The sharks coursed closer, carving past the edge of the ice-dweller. They didn't submerge at all, and Ste Etienne identified bladders worked into the weave of their exoskeletons. Like the dweller, they were things of the surface layer, not the depths.

They then started climbing up onto the ice. By this time the water seemed full of them, executing a weirdly graceful circling dance that somehow never resulted in any of them getting in one another's way. They had extending, clutching limbs hidden within their bodies, which could haul them swiftly onto the ice surface. I braced myself for a battle royale with our escort.

Yet it didn't happen. We were betrayed. Before we even understood what was happening, the Shrouded, who had loyally saved us from harm so often, shunted themselves behind and underneath the pod, and tipped us into the sea.

3.3 DARKNESS

Long before we reached the ice, my understanding had condensed to merely the knowledge that the Stranger has a destination. I have not enough and just enough of myself to not know what it wants, but to know that I am intended to go with it. Every time it rouses from its still moments, it finds the straightest line its limited ability permits, and holds to it. When it sets off, I compare its course to the shifting thoughts of the world, and it shows that the Stranger, too, guides itself by some similar beacon. Its path is constant, each time it moves.

And then the ice. I have fragmentary memories of moving onto this ice long ago. Foraging in hard times, devising hands that can hook prey from the sea. It is not my natural place, though. Yet the Stranger's course demands they cross it and, after a little hesitation, they do so, and I follow.

My following comes less readily, and I retain just enough of my own knowing to despair. The signals that I – when *I* was greater and understood more – painted onto the Stranger are losing their potency. Without them it becomes hard to remind myself why I am following the Stranger and guarding it from the hazards it seems incapable of grasping. Hard, because I no longer know. I only remember this is what I decided must be done, back when I had a greater picture at my disposal. I have a sense of duty, but not the reason for it. And it fades as we go on. I have been this limited *I* for too long, and new experiences are shunting the old ones into silence. All I have is a memory of a memory of dedicating these segments of *me* to following the Stranger. That, and the fading symbols I can barely read. As they fade, my ability to know them fades too.

There will come a time, quite soon I think, when I will not know what to do with the Stranger. Perhaps it will not be prey. It will remain a thing of interest to me, though I will no longer even know that I once knew why. But I will lose that interest. The binding of my mission, self-imposed in that past time when I knew more, will loosen and drop away. And I . . .

I will find somewhere, some hollow to crouch in. I will hunt and feed, dig into the ground until I have made a chamber where my thoughts can sound clearly between my segments. I shall become a new *I*, and the greater I who sent me on this quest will just be one more Otherlike mind to be traded with, to send inscribed and mute ambassadors to.

And the Stranger?

The Stranger will die, I suppose. It is so poorly shaped and clumsy. Swift in its movements, but ignorant of the world. It does not even understand not to pause too long on the ice, and must be saved once again. It never deviates from its obsessive course, even when a great devourer tracks it from beneath, attracted by the dull regularity of its footfalls. I see the telltale thinning of the supporting mass beneath the ice, reacting to the devourer's voice, and I pitch my own voice against it, fighting for influence until the devourer gives up and goes to hunt easier, seabound prey. The Stranger just stomps on, not even slowing.

At least, after that first time, it has not attempted to pause on the ice any more. Its habit of stopping for no reason was just one more confusing aspect of its behaviour that I lack the perspective to understand.

It moves on, and I move with it. Just when I grow hungry, there is some good hunting. Swift ice beasts, carried by the wind so rapidly I must engage my inner reserves even to perceive them. The wind and their own speed are a trap, though. Once I can detect them through their voices bouncing from the ice and their vibrations, I place myself in a cordon in their way, bringing them down in great tangles of limbs and sails. Prised open, their flesh is nutritious, filled with encrypted energy reserves that replenish my own dwindling supplies. The Stranger eats nothing, and ignores any offering of food I make. Just one more strange thing about it.

At the edge of the ice, even the Stranger seems baffled. Helpless as always, at first, so I have to catch it when its weight breaks the ice. But then it cannot just proceed along its straight course any more. Eventually it moves along the ice boundary, rather than simply descending into the water as I would have expected. I may have only a limited perspective on my place in the world, and why I am on this journey, but my voice and the reach of my thoughts tell me the sea is not deep, and the shore is not far. It is the life on the far shore that keeps the ice-bearer away, because they have found ways to harvest it, or they send signals that discomfort it. They value their open water. But the Stranger does not understand this, or perhaps has its own reasons for not wishing to encounter the far shore. Instead it treks off along the line of the ice.

Soon after, the first faint idea intrudes into my mind. It's a fleeting and momentary spark, without context. Losing itself in the interplay between my segments, and between my thoughts as they are sent out into the world, then as they hurry back to me bearing the information the world has for me. A spark, a glitter. Very faint, and yet so much richer than the dull *unh unh unh* of the Stranger.

Normally I would retreat. If I was foraging or hunting, aiming to return to that greater *I* with my gains, this would mark the limit of my range. Because beyond here is an Otherlike, something akin to me, and which perhaps *was* me long ago, before becoming separated and building its own home. But I am not just foraging. I have been sent away from my own home to become nomadic, following the Stranger. I am left with a hard decision, without being left with enough perspective to aid me in weighing up my options.

I could turn back, but the Stranger would not. And even if I did, I do not think I would retain sufficient knowledge of my purpose to return to my former home and resume the life of the *I* that I once was.

Or I can go on. There is a lot which attracts me to that. I desperately need to understand my purpose here, before it all sloughs away and I become merely a stunted solitary self, building up from nothing, with no knowledge of why. I hold onto some decaying memory that I – this I sent out to follow the Stranger – was once a greater thing, but there were threats or fights and segments were lost.

I move on as the Stranger does. I think of the far shore, and how they hunt there. The innovations and adaptations demanded by living in the sea caves. The last faint traces of the signals the Stranger was painted with attract my attention and I consider it. I have a reason to follow its path and ward it from harm, even though I don't know what that reason is, and this still interests me.

I have just about held onto the knowledge that the Stranger wishes to cross the sea, and now I can pass it back and forth and examine it from a greater perspective. That interests me too.

I turn over this knowledge that I have brought to myself about the Stranger, aware that it is only a fraction of what I knew before. Those fading scent-marks reinforce the fact that this odd thing is singular and significant. And that especially interests me.

Many things in the world interest me, and everything that does has been studied and made to serve me in one way or another. I am good at learning, living here beside the sea, because the sea is not a quiet neighbour. Its ways have broadened my mind and allowed me to grow complex and numerous here on the shore.

Now I have come bearing a fascinating gift. Something ill-understood, that behaves – as I see, turning over the piecemeal recollections that accompany it – both with intellect and stupidity. A curious mix. A fading association with the sky, which is surely impossible for so heavy and ungainly a thing. But something from *elsewhere*, definitely.

From my positions within the water, as well as those inherited segments of me that are its escort, I examine the Stranger further. Living here in this liminal, littoral place I have experienced more of life than most Otherlike minds. Whilst the Stranger is unlike anything of the air I have ever been able to pull down for study, the air reaches further than the most aggressive thought I can conjure. So perhaps it really is from some unthinkable realm beyond my imagination. I am a great mind, after all. It takes a perspective as broad as mine to acknowledge the possibility of things beyond grasp.

I know which direction the Stranger would prefer to travel in, and that road will take it to my home here, where I might bring the full circle of my perspective to bear on it. I will, therefore, assist it in its journey.

My seagoing segments stretch out from my home. They are not a roving pack, which might lose my edicts and their focus, but a long chain of many parts, each holding close enough to the next, and the last, so my thoughts can extend across the intervening water. I forge a solid chain, letting the dissident and different memories whirl and osmose into the greater complexity of my mind. Some ideas are diluted and lost, others caught up, reinforced and set down as memories. There is always loss of signal amongst the noise, in these meetings. It is hard to keep hold of so many new thoughts. But I retain enough, and I salvage all of my understanding of the Stranger. I register the elusive charge that I (the Otherlike *I* which sent me out here) has passed on to myself, and that I (the roving, disintegrating *I* which followed those commands to escort the Stranger) brought to me here (the *I* which dwells on these near shores of the ice sea). A chain of self to self to self, all of which were *I* until becoming just a memory in a greater *me*. I catch these tumbling links of chain before they can be devoured by the great silence of lost thoughts.

I call out to the Stranger, inviting it to come with me into the water, so I may take it home. As expected – cued by my past experiences with it – it does not, probably cannot, respond. So I force the issue. It has shown many times it does not know what's good for it. I displace it into the water, then hook it, gathering around its surprising weight. It is, perhaps, the densest thing I have ever encountered, other than weights of solid metal. However, with sufficient hooks and segments, I can move it readily enough through the water. It neither assists nor struggles, just hangs there as a dead weight, bleeding out many different thoughts, none of which mean anything. An uncomfortable thing to be around, like a defective segment babbling, but fascinating all the same.

I haul it into the submerged entrance to my caves and up into the air. In one of the larger chambers, that I have designed to amplify and order my thoughts, I set it down and consider just what it is I have brought to myself.

3.4 LIGHT

'We're out,' Ste Etienne said. For a moment I didn't believe her, because the cameras were still dead and the screens dark. Only the little constellation of indicator lights within the pod's interior staved off utter oblivion.

'How can you even tell?' I demanded.

'Pressure gauge.' Shroud's sea-level atmospheric pressure was like being two hundred metres deep in Earth's oceans, but Shroud's pressure underwater . . .

'I'm glad you built the pod's walls with a bit of redundancy,' I said, trying for upbeat but it coming out a little hysterical.

'Not the pressure I'm worried about,' Ste Etienne said. I could see from my board that she was running damage diagnostics. There were a lot of warnings, mostly for external systems. 'Couple of cameras down. One of the legs is frozen. Damn.'

'You didn't waterproof this thing?'

'Oh, now she turns on me.' Ste Etienne sounded grimly amused. 'Not even the water. It's the alkalinity. We're proofed as much as I could, but it's gotten at some of the more corrodible components, like I feared. We just introduced some trace levels of aluminium into the water table, and I hope it poisons the bastards, frankly. Okay, if I work on the cameras and the other delicate bits, can you clear the leg joints with an arm and a scourer?'

'Blind?'

'An arm with a scourer and a microcamera?'

'Fine.' So I set to work. A pen-lamp and a zoomed-in view of the offending limb's joints became my whole world for the next ten minutes. It wasn't just alkali corrosion from our dip in the ocean. There was a lot of organic detritus clogging up the joint, and that made me worry for all the other moving parts of this machine we were entirely dependent on. It wouldn't take many failures to leave us even more trapped and stuck on this planet than our existing baseline levels of doomed.

By the time I'd finished, a couple of the cameras had been fixed, Ste Etienne using the pod's other arms with a professional ease I could only envy. Then she found where the power to the lamps had shorted out, and repaired that too. We were finally able to see where we'd come to.

Underground, like before. In a chamber where the walls were lined with a gleaming pattern of metal strips, set in curling arabesques. These were eye-leading, save that ours were literally the only eyes to be led. A vertiginous thought, to know that absolutely everything you saw, you were the first and only living things to *see* it. The perspective we had on Shroud was one denied to each and every one of its native inhabitants.

And these inhabitants were definitely on show here. At first it seemed like a menagerie of different species. We could spot some of our treacherous escort, but there were bigger and smaller creatures too, and several of the aquatic forms which we'd been thrown to, and which had presumably manhandled us in here. They were lying like beached seals downslope of us. Their back ends showed alternating left-right fins, while their fronts were bristling with a diminishing spiral of arms that was entirely familiar. They were not a different species at all, these ersatz sharks.

I checked our radio reception for confirmation, and received it in spades. Everything around us was broadcasting on that communication band we'd identified, using the same artificially complicated signal encryption. And busily, too. With only a couple of cameras working, I couldn't count up exactly how many beasties we were in the middle of, but the chamber looked big, and there must have been over a hundred, easily. The enormous spiking traffic on that channel suggested that every one of them was broadcasting all the time, to each other. It was an absolute wall of signal and noise out there. I tried out our little three-beat identifier but it was lost instantly in the radio chaos. Except, when I parsed through the signal traffic immediately following it, I was actually able to pick out the echo, three beats back, and it felt a little patronizing in amongst all that chatter.

At their core, they were all the same species I assumed – the same worm-like thing we'd caught brief glimpses of before. We should have been ready for that, really. We'd seen evidence that their iron basketwork exteriors weren't just dumbly secreted like a snail's shell or a crab's carapace but adapted to their needs. The tool that had so handily opened up Bartokh's pod, in fact, because there was no way in hell they'd *evolved* that naturally. They'd come into contact with our initial fleet of drones, and analysed the fittings, then made themselves hands that would *fit*. They had, essentially, analysed a positive thing, a projection, and conjectured what would need to go in the negative space around it, to manipulate it. That was some high-level thinking.

Or it would be, in a human. Perhaps it was just the kind of problem-solving these things were particularly well set-up to perform. Obviously they had evolved to tailor their exteriors for whatever tasks they needed to achieve, as well as their particular environment on Shroud. Our xenobiology primers cautioned against imputing actual intelligence to any aliens we might discover, merely on account of complicated things they may be able to do. Of course, Concern doctrine was very focused on resource extraction and exploitation of whatever we encountered in the galaxy. Intelligent aliens might introduce a level of moral complexity that would compromise operational efficiency. Cynical of me to think about it in that way, I know, but there was always a good commercial use-case for minimizing the potential rights of whoever's environment you were destroying.

Another screen lit up, and Ste Etienne made a grunt of satisfaction. Personally I didn't think the better view warranted it. The Shrouded out there were constantly in stilting motion, with thirty or so circling the pod, but way more in the gloom beyond. All of them screaming away on their particular set of frequencies, as well as regular audible screaming too. The thick pod walls muted it, but the background thunder and buzz of just being on Shroud was decidedly louder in here.

And where was here? I focused on the one wall of this big space that our light reached. It wasn't like the bored rock of the last nest we'd found ourselves in, of unfond memory. It looked almost like poured concrete, complete with ripple marks. Or maybe clay. Clay slapped on clay while it was still wet, with the surface bowed and curving. Organic, perhaps, and minutely stratified. I thought of reefs, coral building on coral.

Environmental readouts told me it was a good seven degrees warmer in here than it had been in the other nest, which itself had been around the same amount warmer than the actual outside air. The atmospheric composition was a little different too, with levels of carbon dioxide we hadn't run into before.

'Leg done?' Ste Etienne asked me.

'Hm? Oh, yes, sorry.' I'd finished, but not ticked it off on the maintenance list. By now she'd restored all our cameras, although most of the other views were just creatures and darkness, because the space we were in was so huge.

'They've not opened us up yet,' she noted.

'Maybe they're off making spanners.'

She made a thoughtful sound. 'So you reckon that first lot were just lucky – they had a drone land near them and learned to take it apart, then they got us?'

'Maybe.' I knew where she was going, but I wouldn't go there with her. It was a step too far.

'Only, this is a big world and we didn't exactly saturate the surface with drones,' Ste Etienne plodded on.

'We don't know—'

'So maybe one of them cracked it and told the other nests.'

I closed my eyes. I didn't want to have this discussion, since it wasn't one either of us could win. It wasn't an engineering problem with a practical solution. It stepped into the realms of philosophy pretty quickly, because it was all just speculation about the unknown, based on close to zero facts.

'How would they do that?' I asked, even so.

'Radio. They could be pinging ideas around the world in the blink of an eye.' She stopped herself before I could. 'Or no, not with everything else blasting away on all channels, I guess. But if they're smart, they could send the word around. It's more likely than us just happening to end up exactly where a drone had been.'

'You said yourself, wind patterns. We might have ended up on Shroud's one and only drone graveyard.'

'I don't believe that. Again, what's more likely?'

'We can't *know*,' I cut her off firmly. 'And we can't speculate probabilities because there are too many unknowns. We'll go mad that way.'

Ste Etienne lapsed into silence for a moment, perhaps taken aback by my vehemence. In truth I felt a chasm at our feet. A well of fruitless speculation that would absolutely consume us. I didn't want to think about it.

'Let's just deal with what we can touch and be sure of,' I said. 'Which is precious little. We need to focus on making it out of this cave, and to the pole. Speculating about the inner life of these bloody things isn't going to help us.'

'Yeah, well,' Ste Etienne said unhappily. 'I've got some good news and some bad news on that front. Bad news first.'

'What?'

'I was taking a look at your nav data, on the trek over the ice. I mean, full marks on planet-spotting, and especially catching that pissant sliver of dawn. Gives us a heading and we can refine where we go with it as we get closer. Brilliant. Except that *sea* we ran into back there raises the fact we have basically zero information on what's between us and there. Volcanoes, impenetrable forests, chasms, mountains. We need our walk to be as short as possible, while avoiding the worst country. Which means we need to have a more exact course and heading. We're looking at severe wear and tear, and the moment more than one leg gives up the ghost, we are absolutely stuck. We reach some serious mountains or something, that's going to mean taking the long way round, and we won't know which is the shorter long way. Lots of vulcanism on Shroud. Lots of mountains.'

'That's the bad news?'

'That's the bad news,' she confirmed.

'That is bad,' I agreed, fighting to keep my voice steady. 'I recall something about good news.'

'Yeah, well, qualified good news,' Ste Etienne admitted. 'I reckon I can give us some idea of what's ahead. Nav and terrain data both. Using a drone.'

'You've found one of the drones?' This felt a bit rich, after her talk about how unlikely it was we'd ended up anywhere near one.

'These pods come with a lot of toys. Or they were supposed to, 'cept I never actually managed to fit most of them in. But there's a deployable air drone. Kind of micro-dirigible thing. If we climb up somewhere high and open, I can inflate it and kick it off into the airstream.'

'And then? We can't exactly remote pilot anything here. There's the whole problem we had with the original drones, right?' I pointed out.

'Well, that's where you come in. I'm going to bring the physical systems online, and you're going to write instructions for it to go gather nav data and then find its way back to us using the data it's gathered. You can apply our best-guess weather cell models, and I can kit it out with some solid steering fins, so it won't just be battered about like a balloon. Assuming some damn sky-monster doesn't eat it or something. But if we can get it to go up there, and then to come back, hopefully it'll have enough data for us to triangulate where we are and how we make it to where we're going.' She was being fiercely positive now, the engineer for whom nothing was too great a challenge.

'Sure,' I said, not really feeling it. 'Let's do it.'

'Okay.' I saw from my readouts she was putting the repaired leg through its paces, ensuring I'd cleared all the joints properly. 'Good work. Now let's see how our hosts react if we go for a stroll.'

They certainly reacted, but at the same time didn't actually try to obstruct us. We were acutely aware that they'd have been more than capable of fencing us in if they wanted to, given there were so many of them and they were bigger than us, if not heavier. Instead, they just flowed around us as we walked, the whole mass of them moving as readily as a shoal of fish or birds in flight, never jostling one another, never stepping on each other's toes. I kept an eye on radio activity levels, and I almost thought our every move had a shadow in their constant exchanges. As though I was seeing tides of gossip wash back and forth between them. Save that it wasn't anything as inefficient as sound, like the chatter of a mob of workers in the canteen, say: it was at the speed of electromagnetic waves, swift as the firing of neurons.

I like to feel I was teetering on the brink of revelation then, except at this point Ste Etienne piloted us out of the chamber and we finally saw where we were.

3.5 DARKNESS

When I have it home, I consider what it actually is, but am not sure. A fading memory tells me this thing is . . . complicated. That is mostly what occurs to me as I examine the Stranger. A twitch in my thoughts dissuades me from rash action, when I might simply have peeled the thing to find out what it's made of. This stems from a faint legacy of the ideas I inherited when I received the Stranger, and which are now mostly lost within the great bustle of the rest of me. Whichever Otherlike mind sent me these thoughts, there was no time to write a proper message and prepare an ambassador. No cogent and well-structured train of chemical inscription to inform me of just what this thing is. Only a scatter of loose associations, born from whatever events have happened around this thing since first it was encountered. It's vexing, that any concrete understanding has been lost in the process of bringing this prize to me. And yet . . . *complicated*. When I reach out to assuage my natural curiosity in an act of deconstruction, the nagging thought holds me back once more. And so I observe.

It lives, because it thinks. An undirected chaos of thought, crying out in many voices, as though countless different beasts are trapped within it. At first that's all it does. My own thoughts explore the exterior of it: bright, sharp, keen. My voice wraps about it and finds unexpected curves and angles, like nothing of land, sea or air that I can remember. It's a relief when it begins to move. My curiosity is well and truly engaged and if it hadn't begun to perform for me, I might have had to let myself be reckless.

It spends some time cleaning itself and restoring its shell. The actual processes it uses are unfamiliar, but the intent seems clear. I am left with the impression that being underwater has discommoded it more than I expected. The sea is a duller place than the air, but life exists there. My own segments can live and hunt within the ocean for some time, so long as I have fashioned appropriate bodies for myself. And the segments themselves need sufficient internal reserves to endure the lack of sustaining atmosphere. I let my thoughts and voice focus on the fine detail of what the Stranger is doing, as its limbs fidget across its body. Minute parts of it are cleaned away, and I detect evidence of interesting chemical reactions. A little analysis suggests the very substance the Stranger is fashioned of has reacted poorly with the water, and I devote some resources to considering the precise logistics of that. These are not chemical reactions I would expect to encounter, although now I've seen them, I could isolate the relevant materials and replicate them. Simply put, some of the materials which make up the Stranger are not those which would persist in the wild in their pure state. Which is fascinating.

I draw in more of my consideration, letting a certain amount of dull routine activity fall still. Wells of memory open to me, old strata of experience set down long ago, and seldom tapped into. Day to day, life does not require such longitudinal perspective, after all. My deep history erodes and atrophies from disuse. It is only by innovating that I can maintain the sharpness of my mind here. Innovating and occasionally encountering the truly unexpected, like now.

When it starts to move, a variety of my theories concerning its physical form are confirmed. Namely, it does not move like a living thing, or at least no living thing I have encountered. And I believe I have a greater perspective on the world than most, with my home here at the very border of land and sea, with the heights of the funnels as well as the ocean floor available to me. I begin to burn reserves, inflaming my thoughts, giving them speed, allowing me to process greater loads of information more rapidly. The relative motion of the Stranger decreases from a staccato rapidness to a plodding languor. I watch its limbs, the disconcerting symmetry of it, and the rigidness of its single body section. I observe the way its legs and arms are set on wheels which rotate freely about its spherical main-section. I could conceivably construct a body like this for my segments, but it seems hugely inefficient.

My voice returns from the Stranger with a suggestion of hollowness. Considering what might be inside has my segments mobilizing with hooks and levers, but that memory-trace forestalls me again. What is inside is not profitable to expose. That is the understanding rising up to me from the dissolved thoughts I inherited along with the Stranger. Be patient. Observe, but do not touch.

I can be patient, but I can be proactive too. I balance at the midpoint between these desires, feeling different cycles of thought within me come down on one side or the other. The fate of the Stranger hangs in the balance as I consider.

Patience it is.

3.6 LIGHT

We made it outside, to then find ourselves up a mountain range, abruptly teetering on the brink of a drop that would have bounced and rolled us down a warty slope, and into water choked with grinding scales of fractured ice. Ste Etienne swore and clamped all four legs down, anchoring us as best she could.

There were bridges and ramps, we noticed. Structures made of – well I thought it was intricate girder-work at first, but in fact it was bodies. Or, no, that was too morbid. It was empty exoskeletons locked together, limb to limb – welded, almost. A whole mortuary infrastructure. And we had no way of knowing whether the occupants of those discarded shells had died in situ, or been allowed to wriggle out and go make themselves a new suit of clothes.

What those bridges and ramps were connecting were . . . chimneys. *Volcanoes* initially came to mind, because some of them were actually smoking, but they weren't that. Or they weren't *just* that. We worked out from chemical analysis that some sort of vulcanism was at the root of what we were seeing, certainly. Specifically, ocean-floor vents, way down below us, were churning out a wealth of chemicals that the life of Shroud obviously relished. There were communities in Earth's deep oceans that built up around such vents, relatively separate from the rest of the world and its ecological tribulations. On Shroud, I guessed, the active vents didn't migrate as rapidly as their Earth equivalents, because some kind of community had been living off them here long enough to grow enormous. We'd come out partway up one such chimney, presumably not a smoker given the clarity – relatively! – of the air inside it. Our lamps showed us a dozen smaller vents, some of which were discharging visibly murkier stuff into the atmosphere, as well as the sloping side of a peak even grander than the one we'd just come out of.

'Right,' Ste Etienne decided. 'That one.'

'You want to climb it?'

'We need to get high, right?'

'It's too steep.'

'We can do it,' she decided. There was an edge to her voice, because we were looking at something entirely unexpected. A landscape, a biome, that none of the drones had seen. Our self-constructed mission was already veering onto the far side of impossible, and we wouldn't survive many more surprises. How many radical shifts of landscape lay between us and our ridiculously distant goal, exactly? A task that only appeared remotely possible if you refused to actually examine the details. But without this goal, we had literally nothing else to do but wait to die.

I forced myself not to think. Sometimes action is all we have, to stave off despair.

Crossing the corpse-bridge was bad enough. It wasn't designed to take a superdense lump like us, and we felt it shiver and give with each of the pod's deliberate steps. The fact the locals insisted on keeping pace with us didn't help, adding even more weight. And they were constantly almost in the way of where our feet needed to go. Except never *quite*. Blind as they were, they were still absolutely aware of our position, and that of each limb of the pod.

We came to the face of the next chimney, looking at a slope that was close to vertical, extending as far as we could see. Even the locals didn't seem keen to scale it, and they were surely better equipped for it than we were. Or they could make themselves so, I supposed. Tailor their woven shells with crampons and carabiners if they wanted to. But, sensibly enough, they didn't, so the stupid attempts were left to us.

We tried the slope a dozen times, but couldn't do it. Three times we almost ended up rolling downslope into the water. We'd have surely cracked ourselves on the way down, and I'm not sure we'd have survived that, given the pressure differential, plus the extra joy the local gravity would have lent to each impact. The slope was too steep, too smooth, and its texture just rounded ripples from whatever living process had built it up.

The locals watched – or at least eyelessly observed – our doomed efforts. If I was hoping they'd build us a stairway to heaven it didn't happen. Instead, eventually, they ran out of patience with our feeble fumbling, and a dozen of them picked us up and carried us inside. Not back into where we'd been, across the bridge, but into a vent in the side of this chimney. We hadn't even noticed it, out of the range of our lamps as it was. There we found out what it was like inside one of the smokers, because this funnel was still very active.

We'd lost enough height with our scrabblings that the extreme reach of our lamps could now catch the water below. Roiling and bubbling with volcanic gas, and not a trace of ice. The air in here was even more opaque than Shroud standard, the temperature up another ten degrees, actually past zero centigrade. Atmospheric telltales registered enormous levels of sulphur compounds, methane and carbon dioxide, almost an entirely different atmosphere to the outside. And something in here loved it.

Where our light reached the interior walls of the chimney, we saw life. A weird yellow fronded life, that thrust pulsating fans out into the column of chemistry-heavy, heated gas churning up out of the water. I thought it was a host of individual plants at first. Then I saw that the walls were covered with interlacing mycelia-looking strands, and so guessed the entire thing could be one enormous creature. Something like the dweller under the ice. It lined every inch of the walls, and the centre of the chimney was a vast tangled forest of its interwoven fans sieving the thick air. The locals, moving up and down the chimney on a spiralling ramp, were harvesting it.

They had sets of pincers, scissors, shears and secateurs of different sizes, and we watched them pruning and snipping away at the fans, stowing what they cut within their hollow basket bodies. By then we were trudging up their circling path, and I was numbly sure that the Shrouded had finally realized we wanted to go *up*, but that we hadn't been aware they'd constructed an accessibility ramp inside this chimney.

It was Ste Etienne who registered that the mephitic air inside the chimney was almost certainly not what the Shrouded needed for their regular metabolism. The busy workers around us, which always moved aside as we approached without ever needing to see us, were fitted with bladders, just like the shark-forms we'd seen before. Except these were presumably not only lifejackets in case they fell into the boiling water below. They were, we decided, aqualungs. One more example of how versatile the species was. One more bone to throw into the pit of possible alien intelligence, for the competing theories to fight over.

'You ever see the ant farms, in the habitat tanks?' Ste Etienne asked, as we climbed.

'I mean, sure. In passing.' Childhood memories came to me of green walls under artificial light, growing a dense mesh of complementary food crops that didn't need human tending. You'd see the ants skitter about with the constant bustle of insect life. The minute custodians we'd suborned to do our will, and save us the need for our own labour.

'I did my first professional shift on them,' she told me. 'Saw up close how they worked. Real close, with the ants. You know, what we've got them doing, that's not invented from genetic whole cloth, right? It's just adapting existing ant behaviour, making it work for us. Ants don't think, but there's a whole lot of real complex stuff they do even so. This sort of thing here, it's just ants writ large, right?'

I made a noncommittal noise. I wasn't sure if she was trying to convince me or herself. At least this all fit within regular Concern thought. Nothing that would have troubled Opportunities if we reported it. I could tell she didn't believe it, though. She'd decided there was more going on with the Shrouded than brute instinct could account for. But neither of us wanted to put anything on the record.

'It's not the vertical farm,' I said at last. 'They understood to show us the accessible way up.'

Now our positions were reversed and it was *me* arguing for their smarts, and Mai talking them down.

'We don't know that's what they're doing, though,' Mai

said. 'Maybe they're just real proud of this farm thing they invented and wanted to show it off.'

As the pod plodded up the ramp, the worker Shrouded continued to just step aside for us. And whilst they had nothing with which to look back and notice us, and doubtless they were being informed of our approach by vibration, sound, EM activity, scent, all of the above, it still felt weirdly automatic, as though they just *knew*. I'd watched ants. Like Ste Etienne said, the ant-tended agricultural walls were a staple of the habitat tanks, with food and atmosphere regen all in one. Ants are messy. They're constantly bumping into and crawling over one another, stopping to twiddle their feelers and check the ID of the ants around them. But the Shrouded didn't do any of that. They moved as neatly as the pieces of a machine, or a swarm of linked drones.

They were touching feelers, of course, but their feelers were invisible to us. All that EM chatter on the band of frequencies they used, bouncing between them and from the walls. Like most standard-grade Concern kids, I remembered the habitat tanks keenly. They packed us in tight, in those tanks. You weren't given any space to yourself until you were trained and had earned some wage-worth. Small wonder we were all desperate to be sent on the space missions, where you could gain a little autonomy and elbow room. On the occasions when they woke you out of hibernation, anyway. But life in the tanks meant living elbow to elbow, all the time. With everyone's noise and smells and antisocial habits. I reckoned being a Shrouded must be just the same. A hundred near neighbours constantly jabbering at you over your integral radio receiver, constantly telling you . . . what? Was there Shrouded gossip? Social species evolved to compete, deceive and outdo each other, and bitch about each other too, probably. This was a big driver of cognitive evolution, so ran the theory. And just because none of the actual body language I saw suggested this kind of division, backbiting and healthy social Darwinism, that didn't mean it wasn't going on in spades on the EM band.

'Damn me, how far up does this thing go?' Ste Etienne asked, and I replied automatically, 'We're near the top.'

She let that go for about four long seconds, then asked, 'And how the hell, exactly?'

'Atmospheric concentrations of sulphur and carbon dioxide are way down, and the plant stuff, it's thinning out.' I was actually quite surprised by my own perspicacity. While I'd been speculating about the Shrouded, some part of my mind had been on the job too.

'If this just leads to a little balcony right at the top and no way outside, I'm going to be pissed,' decided Ste Etienne. It did, and then it didn't.

The ramp literally just petered out into nothing while we were still inside the chimney. The Shrouded didn't go in for safety railings either, so if we hadn't had a good lamp pointed forwards, we'd have marched merrily off into thin air, falling down a hundred-metre drop into the volcanic water below. Was this it for the great vertical farm excursion? Except the Shrouded then started busily carving us an exit through the wall. They had our size down, too. The circular bore they made in the side of the funnel wasn't quite a cartoon hole the exact shape of our pod, but it wasn't far off. I guess that's what sonar can do for you, if you're precise enough about it.

Ste Etienne grunted. She did that a lot and I was growing

familiar with her range. This one was 'surprised and a bit impressed, but not wanting to admit it'.

Cautiously, she jockeyed the pod through the tight hole the locals had made. There wasn't much footing on the outside of it. What there were, were Shrouded. A handful of them, walking their enormous metallic cage-bodies up and down what was just about a sheer surface. Each of their deliberate steps anchored to the stony stuff with a hooked claw. They were all-terrain monstrosities, given the chance to adapt their footwear.

We ended up clinging onto the hole with all four feet and most of our arms. The drone's release might have a bit of a Newtonian kick to it, and we didn't want to end up in a freefall descent either inside or outside the funnel.

We were high enough up here that, aside from the funnel itself, there was nothing to be seen. The dark air kept its secrets. I registered the tug of wind and cross-referenced our best plan of the weather cell patterns.

'Is it not just going to be blown away and we'll never see it again?' I asked.

'Maybe,' Ste Etienne admitted. 'Or maybe it whacks into a mountain. I mean, we have this one shot, and we're up as high as we're going to get. The drone will be caught by the wind. And the wind is going in the right direction, mostly on account of the fact we're pointed towards the colder half of the moon, the outside-facing half. So the drone sods off that way, and lets itself be carried. I've fitted a trailing radio buoy to it, so it can capture some kind of terrain data, assuming it ends up low enough. I've programmed its EM receptors to take readings on the solar and planetary EM activity related to where it passes through, so we can then triangulate our destination better. It's all jury-rigged to hell. What else is there? And then . . . ?'

'Well, okay, here's my working.' I passed her my instructions, and then the automatic summary. The too-long, didn't-read piece you always provided, because busy Opportunities staff can't be expected to actually go through line by line. I'd given the drone the best nested priority tree I could, which would trigger a return protocol to make it drop out of the wind cell and come back to us, using its own nav data as a guide. We'd have a beacon going, the most powerful signal we could generate. That wouldn't count for much, though, because there literally wasn't a clear frequency across the whole spectrum. But if the drone came close enough, it might just hear us and be able to home in across that last little span of distance. Or at least signal us as it threw itself into a controlled crash.

That was the theory. I'd done my best. Ste Etienne went over my figures – skimming the code itself as well as just the summary – and couldn't find fault. Jerennian would probably have done a better job, but again, I was the understudy, second best at everything. Simultaneously the most and least useful person for Ste Etienne to be stuck in here with.

'Right then,' she decided. 'Let's do this.'

From a camera on the end of an arm, I was able to watch a circular hatch pop open and the drone appear. A little like watching a butterfly emerge from a chrysalis, really. The hard parts came out disjointed, loose-linked, then snapped into rigid place once it was in the open. A plastic lozenge, about the size of a human torso, clamped to our hull and extended a bristle of antennae. Its skeletal wing framework clicked into place, section by section, and we ran each electric engine and propellor in turn, then had the drone's onboard computer do the same, so we could be sure it was all linked up.

The balloon began to inflate, sucking in thick Shroud air and expelling everything except the hydrogen, which formed such a sizable fraction of it. We'd looked at the logistics of dirigibles on Shroud from the safety of orbit. The results had been mixed. The high atmospheric pressure meant you needed to cram your hydrogen in there to keep the balloon pushed out and inflated. But the same universal access to atmospheric hydrogen meant that hydrogen wasn't actually as good a lifter as it would have been on Earth, because there was so much loose in the atmosphere already, as an ongoing by-product of Shroud biochemistry. Despite all these strikes against, though, it had apparently been a more efficient way to design a long-range pocket drone than relying on heavier-than-air methods. You couldn't beat it for battery and weight efficiency.

After a while, the balloon – made up of multiple hydrogen capsules within a single self-sealing skin, in case of puncture – had become large enough for Ste Etienne to cautiously deactivate the magnetic locks which held the drone to us. It bobbed sluggishly free, still tethered. She then ran through a series of tests – once the device was loose, it was on its own. All we could do was give our surrogate its best shot by making sure everything was working before it left us. The envelope continued to inflate by degrees, the drone slowly lifting away from us, then jerking further upwards in the stream of warmer air from the chimney. Ste Etienne directed it to lower its radio buoy, a fist-sized ball on the end of a tether, which she hoped would hang close enough to the ground to gather terrain data. Given the electromagnetic conditions on Shroud, it was all horribly tenuous.

We finally cut the rope, and watched it ascend gracefully. In seconds our lamps could catch no sign of it. In less than a minute, we had lost its radio trace against the background concatenation.

Then we waited.

3.7 DARKNESS

I watch it as it exits my living chambers, out onto the archipelago exterior, and then awkwardly tries to navigate the infrastructure I have built there. It is trying to reach somewhere. Another thought I have inherited. I feel a brief knot of frustration that so much of this entity's recent history has simply been diluted into uselessness as it came to me. Without the expensive and time-consuming business of a proper ambassador, so much is lost between myself and the Otherlikes.

And the thought comes, that it was not always so.

I have been, after all, plumbing the very depths of my recorded memories. The old, old times, before I came here. Back when the *I* that I was, was a different one. Before the part that would become *me* was sent away and lost its focus, and eventually came here. That was a long time ago, but at the very extreme reach of my recollection, there was a time even *longer* ago when *I* meant something greater still. When things were not so broken up. But there my thoughts end. All I have is the knowledge, shorn of context, that it was so. It was too long ago, and I can no longer piece the disparate parts of the idea together. The one useful piece of memorious salvage I retrieve is that, all the way back to even *that* time, nothing like this Stranger exists in the memories left to me. Nothing that moved, sounded, tasted or occurred to the mind like this. And nothing with such a maelstrom of bleeding thoughts, as though it seeks to communicate with every beast in the world at once. Yet I recall that if I signal it just *so*, let my thoughts reach out to it as though it is *of me*, then it responds to me. Crude and basic, but a coherent response. *Unh unh unh*. Like some strange trick it has learned. Perhaps I taught it this, as part of those experiences that were lost when it came to me.

By now it is trying to climb higher up my domain – a mindless, instinctive scrabble for height. It is lamentably equipped for such progress, though, and lacks the intellect to alter its body and feet for the ascent. For a while I observe its failures, seeing less and less promise in it. Eventually, I guide it inside, where an ascent might be made. I want to discover what it will do, and find out how it will react. This is, after all, one of my innovations. In disturbing my older memories, I have re-experienced how it was when I came to this place to settle. How that lesser me I had become, which was cut off from the greater self I had once been, survived. These shores were not hospitable country back in those days. I had not invented how to live with the sea, how to hunt on the ice, or any of the ideas that have made me the thriving success I now am. I came here because there was heat and life, and there was little left of me. I not only survived, I grew. Idea by idea, I adapted to this new environment, built my home, bred and outfitted more segments, and thus became *more*. And in turn had more complex ideas. I discovered the life which had built the archipelago, and which still lived within it, harvesting the rich air. I cultivated it, and thus became less reliant on sending out so much of myself to hunt and forage. Those dangerous separations, which could often result in whole chunks of myself becoming lost, losing their purpose and never coming back to me, just as I had never gone back to the greater *I*. Now I could feed myself safely at home and grow greater and more complex in my thoughts.

I am, I think, the perfect mind to examine what this Stranger is. I have been honed by hardship, and by opportunity. If this thing can be understood, I will manage it.

Even though it is a single segment, it too can grasp an opportunity. It ascends, and I observe, still. I have an idea what it may be trying to do. It is not like any beast I know, but there are some commonalities across all life, surely. There are only a few reasons why something would aim so fervently for height. The Stranger is gravid, and wishes to reproduce. Releasing young into the high airstreams is a beast's way, of course. I, being intelligent, husband my young so they may swell the ranks of my segments and increase the power of my thoughts. But beasts do not have these considerations. And perhaps, despite all its weird aspects, this Stranger is no more than such a creature.

It signals to me, those mute grunts. I respond, as I remember I did before. Each time I await some more complex train of thought from it, but it seems incapable of anything beyond such simple signals. Sometimes less, sometimes more, most commonly that trio of pulses.

At the apex of its journey it requires the open air again, of course, and I let it out, almost indulgently. By now I am starting to make plans to butcher it as an exploratory exercise, feeling it has exhausted its stock of novel behaviours. In this I am incorrect, however.

When it goes out, it does indeed seem to release a kind of progeny into the world, but not the flurry of minuscule motes I would expect, to be carried by the wind to wherever. To land where they will attach and grow, and become more mindless animals in whatever part of the world the currents take them to. Or become prey to the hunters of the air, or fall into an environment unsuited to them, in the inefficient way of beasts. Except the Stranger has only one child and it is huge. My thoughts reach out, my voice tastes it, finding a solid body which is still a sizable fraction of the Stranger's own. It unfolds with a mechanical ingenuity that shocks me with a dozen new thoughts about how structures might be made. Then, whilst the child remains attached to the parent, joined by weak, whispering thoughts, a greater body grows above it. Hollow, light, a capsule like those my seaborne segments use for flotation in the water, but of sufficient potency that it rises in the air.

I study what it has created and, segment by segment, the routine activity of my home falls still as I am consumed by new thought. I, the great innovator, who took this unpromising locale and made it a great and bustling home, am schooled by this mute, unthinking creature.

The child finally separates from the parent. The wind takes it, and it is gone. I feel a snarl of lost chances. Is that it? Perhaps the Stranger will die now, its life cycle complete, and I can dissect it. Except it is doing yet another new thing. It is crying out, sending a shrill, wordless thought as far as it can, as though suddenly bereft now its child has flown.

I hold off on my final investigation. It seems this is not over. Another stage of the Stranger's drama is yet to play out.

3.8 LIGHT

'They going to wall us in here, you reckon, if we just sit?' Ste Etienne said at last.

'We can move if they try it. It's your turn to sleep,' I told her. We'd eaten a couple of hours of her downtime, in fact, with the whole drone business.

'Not tired,' she said.

'That's what the shots are for.'

Another grunt.

'How long do we wait?' I asked her.

'Your guess's as good as mine. But I figure if it goes forty hours without a peep, it hit a rock in the sky somewhere. Or something ate it. All that nummy hydrogen.'

'Sleep,' I told her.

'If I sleep, can I leave you a checklist to go through? May as well get some maintenance done, if we're here for the duration.'

She'd been busy on the ice trek, making that list. During my solitary time awake, I must have run diagnostics and checks on just about every individual system the pod had. First I checked over the actual damage-control algorithms themselves, and then they and I went over everything else with a fine-toothed comb. I tightened up code, cut redundant loops and other inefficiencies, and polished everything about the way the pod operated. It *shone* when I was finished with it. Digitally shone. But it didn't matter. I knew it, and Ste Etienne had known it when she gave me the list. Because it wasn't defective code that was going to do for us. It was Shroud. It was the physical components of the pod. And it was the malign interactions of the two, over the unconscionably long slog we were blithely assuming we could complete. We'd take a tumble, or a monster would have a serious go at us. A really serious impact might breach the hull, allowing the fist of Shroud's atmosphere to crush us. But it didn't even have to be that severe. Damage the legs enough, or some other vital system, and we'd die just the same. A longer, drawn-out and miserable death, though. We could last weeks, immobile inside our ball. I wondered how long it would be before we hacked the safeties and opened the hatch to the outside if that was the case.

Even if it wasn't some single savage shock carving off a leg, it would be wear and tear, like Ste Etienne had said. Something would give. A joint, a motor, a part we couldn't fix with the limited ability of the arms. And one we couldn't replace or cannibalize from somewhere. It was, I felt, inevitable.

It wasn't a happy solo shift, is what I'm trying to communicate. Not a good time to be left to one's own thoughts. When Ste Etienne woke again, I was almost pathetically glad.

No signal had come from the drone, of course, but then we wouldn't have expected it yet. No hostile action from the Shrouded, either. They were very much still around us, inside and outside the funnel. Every so often the pod had informed me that they'd sent our three-beat signal to us, and I made sure to always reply. Attempts to encourage them to recognize mathematical patterns, through the brute medium of different numbers of pulses, just resulted in parroted responses. The Shrouded could maybe count, but didn't go in for arithmetic. Or possibly they were sending us the most elegant and sophisticated proofs of the nature of the universe. But other than those simple pulsed sequences, their radio chatter was a bewildering storm of noise we had no hope of decoding.

Ste Etienne, brimful of stims and wide awake, looked over my maintenance reports and grunted her satisfaction.

'As good as it's going to get,' she decided, which felt like the epitaph of all our efforts. A desperate, doomed trek to reach the distant pole; as good as it was going to get.

'You may as well put your head down yourself,' she added, after a moment.

'I don't want to screw with my biorhythms.' In truth, I had a few scheduled hours left with both of us on-shift, and even unspeaking human company was better than either being awake alone, or the dreadful dreamless oblivion that the pod's pharmacopeia would send me into. Honestly, neither of us had a biorhythm left standing by then.

'Do you . . .' Ste Etienne's voice had a weird tone to it. Someone who didn't really make small talk, or wouldn't have made it with *me* in any event, but who could tell I needed it right now. 'Do you want to talk about Bartokh?'

'I . . . No.' A stab of loss, although it didn't really feel like the loss of someone I knew – more like the loss of someone I used to be. The life we'd had, up in orbit, after they woke us from the freezer. The halcyon days of Special Projects. A horrible, almost nauseating feeling came over me that maybe those had actually been the best days of my life. If so, how wretched that my life had no better times to show for it than those. 'Why? Do you?'

'I mean, I thought . . .' She was sounding really awkward now. 'Only, we . . . we all assumed you and he . . .'

I blinked, my mind utterly blank. The whole business of being abandoned to die on an alien planet was briefly not a concern at all in the face of this horrendous social embarrassment. 'Me and . . . Oswerry Bartokh?'

'Weren't you?' Ste Etienne's voice was that of a woman not quite knowing what to do with herself. 'I mean, all that time in his office.'

'I was his assistant!'

'Well, that's what they call it.'

'Mai!' I was scandalized, actually shouting at her. Incredulous, horrified, mortified, on the verge of hysterical laughter, all at once. 'What? Seriously, you thought that?'

'I mean, I . . . not *me*, exactly but . . . Okay, yes. Everyone thought that.'

'Wait, did you . . .' There was a whole extra level of ghastly revelation waiting for me. 'Did people think that was *why* I got the post?'

I could imagine her eyes flicking shiftily from side to side, avoiding a gaze that she couldn't in any event have met. 'I mean, sort of.'

'I was good at my job!' I insisted. 'I was good at Bartokh's job! Who do you think did most of the actual work you saw?'

'Okay, okay, I mean I see that *now*.' She scrabbled desperately for a position she could hold. 'Only we all thought it was *him*, back up top. He certainly let us think that, the old bastard.'

I made a wordless sound to indicate precisely what I thought of that, and respect for the dead be damned. 'Well, we all knew you were sleeping with Jerennian,' I told her, as revenge.

She made a choked sound. 'I would never.'

'You don't like Big Mike?'

'Like him, sure. Good worker. Not my type though.' A pause, plainly on the very brink of confession, and why not? What was there to lose, exactly? 'I was buffing Umbar, if you must know.'

'Umbar? Technical Oversight?'

'That's a fancy word for shagging.'

'No, I mean—' I frowned. '*Buffing?*'

'You never heard of that before?' Ste Etienne suddenly very worldly wise. 'When you're putting out for a superior, for extra credit. Like you were doing with Bartokh, or we thought you were.'

'And what was Umbar doing with you then? Roughing it?'

'Resourcing, when it's the other way,' she said, straight off. 'How is it you don't know any of this?'

'I . . . I mean, all right I did know *that* one. I just didn't think . . .' It was short for 'testing the human resources pool', which had been the standard cover for anyone who decided they were going down the wage-worth ladder for their jollies. Bartokh had not been resourcing me, and I had not been buffing him, yet somehow everyone had just believed that was how it was between us, and I'd never realized.

'You did Skien once, though,' Ste Etienne added. 'Or was that not a thing?'

'I . . .' My automatic flare of indignation deflated. 'Well that once, yes. But I think they were imagining Rastomaier.'

She barked a laugh, then wheezed a bit as she recovered from the sudden motion. 'Ow. We are going to be wrecks when they recover us,' she said. 'And no sympathy. I can hear it now. *You spent a month just lying on your back on a padded couch? Back to work, you slackers.*'

My turn to laugh, hollow as it was. 'We are going to be heroes,' I decided. 'They'll write up our story for transmission across the whole of the humanosphere. Survival against all odds. Never give up. Never despair. See what Ste Etienne and Ceelander endured, in their desire to rejoin their crew and become useful workers for their Concern? Our faces on official dispatches, on every screen.'

I was being so damn brave for her. In my head it wasn't even our faces, just our names, and featured in a single terse note in the *Garveneer* personnel record.

I did sleep, soon after. The restorative, blank void of chemically assisted slumber, which I was sure lacked some vital psychological component real sleep would provide. I woke automatically when the pod started moving, because this was what I'd set the system to do.

'You've got another two hours and change,' Ste Etienne told me, preoccupied.

'What are we doing?'

'Your biorhythms, you said.'

'Screw my biorhythms. What's going on?'

'Drone signal.'

'*What?* And you didn't wake me?' I demanded.

'Look,' she said. 'I don't know what there is, other than a real faint ping which could be from the drone. I was planning to wake you when I found it, and actually had it in camera-range. Because right now I might be homing in on some damn monster's mating call or something.'

We were descending the chimney. She'd anchored a line up above and we were performing the universe's slowest rappel down the slope. On every side the Shrouded just stilted past us, keeping us company. They wouldn't be able to catch us if we fell, though. We'd bowl them flat on the way down.

An hour later we were picking our way along a landscape with funnels and chimneys to one side, and a short drop into the sea on the other. The ping was stronger, though, identifying itself as the drone and reporting a variety of system failures. We'd take that. We hadn't exactly expected the thing to come back to us polished up and given a fresh coat of paint by the locals. An hour after that, we arrived at where the drone had come down.

It was a wreck. The rags of the deflated balloon were draped all around it like spider silk, and its casing had one curved corner knocked solidly in where it had struck something. The radio buoy was gone too, its tether ending in a ragged trail of cables.

It was surrounded by Shrouded. They were, after all, faster than us, and if we could hear the beacon then so could they. They hadn't actually got to work on the drone itself, but they'd torn the wing framework off and dismantled it with extreme prejudice, as well as most likely finalizing the demise of the balloon. We rushed in to retrieve the actual data core while they were busy with the accessories. Neither of us said it, but we knew it had been a close-run thing. If we'd arrived to find the Shrouded had wiped our data, I don't know what we'd have done.

There followed some of the tensest minutes of my life

while we tried to actually link to the drone's core and retrieve said data. For a long while it seemed the beacon was literally the only intact part of it. But it had found its way back to us. My own programming, second-rate as it might be, had come through. Which meant something in there had been working as intended.

Ste Etienne finally spliced a link together and the data began to flood in. Meteorological, navigational, terrestrial. All of it was eaten away to crap by the corrosive noise of Shroud, but amenable to clean-up. A ghost of an idea about the world ahead, for us to decode and ponder on the road, like priests consulting the cryptic words of an oracle. The drone was dead, but it had come back to us with an olive branch nonetheless.

Within two hours of electronic haruspicy we had a rough map of what lay ahead. Give or take, within some fairly broad parameters, and subject to official review et cetera et cetera. Furthermore, the nav data from the drone let us narrow down our destination to within, we hoped, a few square kilometres. Close enough that, if we ever got there, we should be able to detect some human activity and home in on it. Or so we hoped. We had triangulated our own current position, too. We weren't far off what we were calling the equator – not the actual rotational equator, but the midline between the inward and outward facing hemispheres of Shroud. We only had half a world to walk over.

All, obviously, cause for much rejoicing. We were practically home.

When we set off again, I watched the Shrouded nervously for their reaction. It was within the bounds of possibility that this new nest of them had grown attached to us. Instead, an escort fell into step with us, just as before. Not even, insofar as I could tell, the same individuals. I reckoned I could tell the design-work of this new nest, and these were locals. We sent our three-beat salute to them, and they returned it readily enough. They learned new tricks quickly.

On the slopes of the funnel archipelago, the wreckage of the drone was soon left far behind.

3.9 DARKNESS

Slowly, the regular activity of my home resumes. The Stranger remains fixed in place high in my domain, letting out its forlorn wail. Sometimes it signals me, and I grunt back. I begin preparing ambassadors to send to the closest Otherlike minds, to inform them of what I have discovered, and to ask if they have encountered anything similar. I have a dim recollection that even though the Stranger is singular, it's not unheralded. There was some prior contact which prepared me for its oddness, though the details have fallen away. The ignorance is maddening.

Eventually, the child it birthed comes back. I know it before the Stranger does, because the child's own thoughts are the same shrill wail, announcing its whereabouts to the whole dangerous world. I then have another sudden reversal of understanding. It is not a child at all, but merely another segment of the Stranger. Sent out as I would dispatch a foraging party, trusting that it will return to me when its task is done.

I race the Stranger to it, a contest I win easily. At the same time, I watch the Stranger lower its bulk down from its high roost, and I garner new ideas about mechanism and design. Everything it does seems ill-suited to its environment. Every solution it has, to the regular problems of movement and existence, is unprecedented. Its own ingenuity is wasted on it. Just as well I'm present to observe and learn.

By the time it arrives to reclaim its segment, I have already studied much of the thing's structure. Studied to destruction, in fact. The Stranger ignores the fascinating outer body, which has given me so many thoughts, and just toys with the hard kernel I was unable to open in the time available to me. That being done, it begins to head away. Those scraps of inherited memory continue to tell me it has a direction it is compelled to move in, and now this appears once again to be true.

I could let it go, to pass out of my life, or I could stop it and examine it more destructively. Although once again that tatter of memory arises in me, suggesting I would ruin more than I would preserve. Or I could do what I dimly recall doing before, and send parts of myself with it, to gather more knowledge which might come back to me. Or to an Otherlike, who would inscribe it for me.

I have learned a great deal from this Stranger already. It has given me so much to think about. Thoughts about new ways of being. New chemistry, new mechanics, new principles of structure and motion.

It would be a shame if it was simply lost.

My deliberations spread out through my home, form cells, debate, agglomerate into factions. They contest, overcome one another, form a decision, all while the Stranger takes a few laborious steps forward.

When it goes, I go with it, a convoy of segments shadowing its movements to the edge of my domain and beyond. I stretch out a chain of myself as far as I can, until I am separated from my greater self. Then I carry on following the Stranger, rehearsing my mission, reminding myself why this is important. Out in the wilds it's easy to lose track of why I'm here. In my midst is the ambassador, though. When the Stranger leads me into the reach of an Otherlike, what I have written in my emissary will explain everything. I hope. I only have a vague grasp of what that writing says, now. But I continue to rehearse my mission: Follow the Stranger. Preserve it. Learn from it. I do my best to hold onto the why of it all, but in the end those thoughts, so complex, are the first to fall away.

INTERLUDE THREE

Let me put this into words you can understand. Life began to scream. At first it screamed into the echo of silence which had first owned the land. Then it screamed to overcome all the other screaming. The world became a very noisy place in so short a time.

Out of all these sounds arose the *sensory voice*. The voice which returns to the ear bringing information on the physical nature of the world. At a range beyond the groping reach of limbs and antennae. With an acuity and informational load beyond the capacity of scent, and far less vulnerable to the environmental disruption of wind and weather. The handful of creature lineages that independently developed this facility became the next masters, and the rest died out, or were driven into specialized forms, eking their living at the margins of the world. Looking back, this change marked the next great age of the world. A mass extinction and replacement, driven by a single sensory innovation.

For a while, evolution was driven by this arms race. Ever more complex sensory apparatus competed for range and detail in the dark. Everything screamed, fighting over the frequencies of sound best suited for perceiving the minutiae of the world via echolocation. Predators fought to pin down their prey, and prey listened out for the sensory signals of their predators, tipped off and fleeing before the voices of their enemies could bounce back to give away their location. Organs of sound projection and reception became more and more sophisticated, a battle of ever greater developmental resources for constantly diminishing returns. Until some species, losing out in the war of senses, invested in other strategies instead. A variety of behavioural and structural modifications arose with the intent of baffling the sonar. Ways of moving that did not register as motion, as well as shapes which seemed like part of the landscape. A hundred ways to avoid notice. And amongst these, one minor lineage grew a net of filaments all across its body, to break up its sonar profile into static. To become a ghost in the world. Because sometimes everyone else is shouting so loudly, you just can't compete.

These filaments, or fine hairs, had other advantages too. They detected electrical fields – a whole new channel of environmental information in the thick, damp air of the low-lying sea-adjacent regions. A channel uncluttered by all the screaming. At the same time, air-dwelling life was expanding across the world, clambering over all the new and complex environments that previous generations had constructed. New environments meant an expanded range of elements in the diet, which would then be secreted out into the increasingly complex suits of mail that life required for support and locomotion. Organic compounds, metals, crystals. Crystals that could generate piezoelectric signals when placed under the stress of, say, being part of a flexing exoskeleton. And signals that can be picked up by a network of fine metallic filaments evolved to baffle echolocation. They do this either directly, one organism to another, or after being projected outwards by a directional network of such filament antennae and bouncing off the surrounding world.

All of life was screaming at each other, louder and louder. A darkness so cluttered by competing sounds that, literally, nothing could hear itself think. Yet one species discovered a whole world of silence it could use to navigate and communicate.

The next age of Shroud began.

PART FOUR
FEW / MANY

4.1 DARKNESS

The memories I have suggest there is hard going in this direction. I do not travel this country. Any more? Again? I cannot be sure. Within the range of my knowledge, at least, I have not returned here. I think it is 'returned'. I cannot wholly rely on that, but it's the feeling I have. Hard country, poor hunting, and my harvesting cannot prosper away from the vents. But also, I recall danger. I have encountered harm here, or segments of me have not returned.

My memories are patchy. When I separated from home, and the long chain of thought finally parted and left me alone, there were other things on my mind. As I was rehearsing my mission regarding the Stranger, things such as information about the world ahead were allowed to fall away.

I cast my thoughts ahead now and they return to me with an understanding of how the land lies. It rises, higher and higher, a creased ridge of old rock, snarled over with the echoes of stationary life. Steep. A dim recollection finds me: the Stranger does not do well with steep terrain.

Other memories arise too, broken pieces breaking the surface and then sinking down. A growing sense of threat. Past encounters with unknown things. Then a complete chain of thought, ragged at both ends: I *did* come this way long ago.

When I was homeless, with my link to my original self gone, and the bounty of the vents was not yet found, this was my road. I crossed these mountains. Not over, but through, within the earth. And that was where the fear was. Where I was diminished, until I forgot everything I ever knew of who I was.

Perhaps the Stranger will turn back, and we can go home. And if we come to the caves, I will try to warn it away. The thought stirs up more ideas, clearer ones, because they were a part of the brief I rehearsed at the point when *I* became this separate I. The Stranger will not understand. I have to try, though.

Underground is a good place to make a home, where the world cannot intrude into my thoughts. But not *this* underground. It is claimed by things that cannot be thought about. It is a place of hungry ghosts.

4.2 LIGHT

'We knew this was here,' Ste Etienne argued. 'The data from the air drone says we can make it. There's a whole course we've got plotted out.'

She was mostly arguing with herself. 'I've seen it,' was my entire contribution. The course wound back and forth up the mountains, curling about foothills, and finally reaching one particular pass which she and the algorithms reckoned the pod would be able to make. Better that, she reasoned, than the enormous detour of circumventing the mountains. The data did not take into account any kind of extra biological impediment the route might invite, though, and we weren't even quite sure of how the weather would impact on us. The windspeeds up at pass level were quite severe.

Ahead of us the lamps showed our path rising. If I'd sat down with the topographical map the aerial drone's buoy had given us, and cross-referenced it to the actual terrain we were seeing, then I'd bet it would have come out accurate within a good tolerance. Eyeballing it, it looked far steeper.

'It's the plan,' Ste Etienne insisted. 'We go around, we're deviating from what the drone saw, and then we're in danger of just going off course, because we'll lose the fine data on where we are. No satellite navigation, remember.'

'Sure.'

So, with what turned out to be my entirely reasonable misgivings, we set off on our originally planned course, ascending into the foothills. The Shrouded followed us as before. My imagination suggested a hesitancy in their movements they hadn't previously shown, but that sort of confirmation bias was what imagination was for.

We had been moving over solid ground ever since leaving the sea – none of the vegetal briar substrate we'd originally come down on. The foothills were hairy with some variant of it, though much smaller in scale, barely coming halfway up the drone's belly. The Shrouded high-stepped through it and didn't seem to have any difficulty. It seriously slowed the pod, though. We had broad feet, and something about the interlaced structure of the stuff closed over them every time we put one down, which meant we were constantly fighting to pull free. It was lost time, and also a strain on all the leg joints with every step. Ste Etienne kept bringing up the diagnostics, watching graphs and bars slowly inch from green towards less happy colours. The ones which were even green to begin with.

There was worse to come. Higher up we started to see evidence of other life. Rather than altitude bringing scarcity, every damn thing on Shroud seemed to be desperate for height in this part of the world. Blessedly, the undergrowth stuff increased in scale, until we were walking under and in between it. We still had to battle, and there were other species that had come to colonize the gaps with stringy stuff like messy spiderwebs, which we wrecked as we burst through. There were cables too. Taut thick strands ascending up into the dark, swaying slightly, and anchored to the rock with clutching holdfasts of tendrils half the size of the entire pod.

Up higher still, we started hitting the stronger windspeeds. The blow was coming from behind us – relatively warm air being carried towards the more frigid regions of Shroud that faced away from the star. It howled and funnelled through the maze of briars in strange ways. It also complicated our steps – a whole extra level of calculation was needed for each one, to stop a sudden blast bowling us forwards. Or off the ledges of rock and living material we were having to cross. To our left, the cameras looked out onto a void. It didn't help that we knew what was there, because we also knew just how far down it was. We'd covered a lot of vertical distance by then, and it was entirely open for us to cover it a second time in the opposite direction and at far greater speed.

There were monsters there, in the abyss. When I first saw one, just looming into the reach of our lamps, I actually screamed. On the basis that Ste Etienne didn't bawl me out about it, she'd been spooked too. We saw a vast maw, crammed with interlocking spines, bending close to us, as though it was about to swallow the pod whole. It was the head of something which must have been hundreds of metres tall. As it surged closer, Ste Etienne locked the feet to the rock and prepared to go down fighting.

We saw the neck then, as the wind drove it so close to us we could have fended it off with a stick. A neck as thin as a human arm, no more than the other end of one of those cables we'd seen lower down below. A kite monster, like a huge living windsock, akin to what the early drone had seen, back when Shroud and its life had been a more distant and theoretical problem from the comfort of a workbay. Buoyed up by the constant cycling winds, it must have been made of honeycomb, incredibly light for its size and possible only because the meteorology was so constant and the atmosphere so dense here. Its maw gaped, but the forest of feathery projections within it was for sieving, not for chomping down on offworld explorers. We relaxed.

Ten minutes later we had the really bad news. We reached the pass.

The radar data from the drone's buoy had said we could make it over, but it hadn't accounted for the locals. The same reasons that made the pass a natural crossing point for us also made it a funnel for the wind, and on Shroud that was peak dining for a great many things. The pass was a kind of basketwork jungle, as countless Shroud creatures fought for space to stand with their mouths open. Our lamps only touched the outermost layers, but we'd have had to hack our way through with whatever the pod carried that most closely resembled a machete. Even our escort would have struggled, I think. And that wasn't taking into account anything which might live within the tangle and be less than delighted about being disturbed. It seemed mad that the equivalent of an alpine habitat was quite so busy, but Shroud's priorities were very different. The air-feeding life we'd seen elsewhere had been relatively spread out, so big things like our friends or the pod could move through it at will. Here, the swift-moving aerial buffet was the equivalent of fertile ground on Earth. Everything wanted to live here.

We spent a while just staring, while small, many-segmented things within the tangle threatened us with splays of limbs like multitools.

Around that time, I also realized our escort had been signalling us for a while, with that three-beat signature tune which had become the single word in our Earth–Shroud dictionary and basically meant, 'Hey!' They kept going off and coming back, or some of them did. And when we finally went with them, we saw they'd found a cave. A big one, too. They were seriously agitated about this cave, weaving about in front of it like bees dancing, and pinging us over and over.

'Don't love this,' Ste Etienne said.

Neither did I. Going into a hole to die wasn't on the bucket list, honestly.

Although . . .

There was a fringe of stuff – hairy stuff – around the cave mouth, and extending inwards as far as the lamps lit up. It flurried constantly, like it was the fur of an inverted cat being stroked by a ghost. Which was the weirdest image that ever popped unbidden into my head.

'There's a way through,' I said.

Ste Etienne grunted interrogatively. All that airflow was going somewhere, though. We waited for a while and there was no great outgoing breath, as from the mouth of a slow-breathing monster. Just a constant passage inwards. Through the mountain.

'So,' I said, 'underground, then?'

'You trust them?' Ste Etienne said. 'I mean, all they did was show us a hole.'

'Well, yes, but on the other hand they did actually show us a hole. They did a whole thing about it. You ever see those rescue-assist dogs they have, the cyborg ones? They made them so they could signal you over comms, when they find people in the wreckage, that sort of business. But even then, they still do the dog thing, you know? They come up to you and whine, then go back to where they want you to be. I swear it was just like that with the Shrouded. Like they really wanted us to know this cave was here.'

'Crazy what passes for entertainment on this planet,' Ste Etienne said. 'I still don't like it.' She sighed. 'Look, I have a single ball drone I can get working, with a bit of spit. Little walker guy. I can fit it with some sonar and a basic maze-solving kit, and send it in. If it makes it to some place at the far end, where its pings find an opening big enough for us, then fine. Okay?'

'Okay.'

We half expected our escort to follow the ball drone in, maybe with the aim of taking it apart like they had the flier, but they just clustered by us as it vanished into the dark of the caves. Although it was no darker in there than everywhere else, I supposed. Which ought to have made me feel better about going in, but really didn't.

We slept. The Shrouded came and went, presumably foraging. They didn't sleep, I noticed. Or maybe parts of their brain-analogues slept while other parts stayed awake, the way Ste Etienne and I were running the pod. But they seemed creatures without downtime, without rhythms. And living without a sun didn't necessarily mean this should be so. They could probably track the star's movement across their horizon, as Shroud's lunar orbit carried the moon into and out of its influence. But, unlike on Earth, any equivalent of 'night' and 'day' wouldn't change conditions on the ground much. There wasn't any span of time when your senses were less useful, or when it was better to hide yourself away. So I slept and Ste Etienne slept, but Shroud and everything on it continued in their wakefulness.

Ste Etienne was already running through maintenance routines when my own couch jolted my eyes open.

'Any word from the ball drone?' I asked. It turned out that, yes there was, and she had just been letting me finish my scheduled rest. Which was simultaneously kind of her and a little bit patronizing. The ball drone had returned after sounding out a way clear through to open space on the other side, in amongst mapping a truly extensive labyrinth of tunnels, most of which were more than large enough for the pod to traverse.

'Which begs the question,' she noted, 'of why, exactly.'

I'd been chewing over that before I slept too. On Earth, caves are mostly water-carved, the result of rain moving through faults in the rock, eroding and dissolving the paths of least resistance. But it hadn't actually rained yet on Shroud. The air was hazy with liquid – the ammonia-methane-water mix of the sea – that was constantly freezing out in a glitter of ice, condensing on the hull to cause us maintenance issues, or being carried up into the higher reaches of the atmosphere. It didn't seem to come down as rain, particularly. The density of the atmosphere merely allowed it to filter down in a kind of constant, semi-frozen haze, which became part of the omnipresent murk our lamps struggled so hard to cut through. It probably still ended up finding its way through the rock and doing some cave-excavating while it did, but the ball drone's data showed caves way too spacious and complex for that. We were looking at some other part of Shroud's ecosystem, cut off from the rich air above.

We were shorn of options, and we were veering back towards feeling we had the measure of Shroud again, which at no point ever turned out to be justified. So: we went in. It was a maze but we had a map. What was the worst that could happen?

The biggest change, entering the caves, wasn't even immediately apparent to us. To the Shrouded, it must have been screamingly obvious though. They were certainly hesitant to follow us inside, and the way they moved definitely changed. They were slower, and not ranging so far ahead of the pod as before. Individuals amongst them kept stopping and having to be chivvied along by their fellows, or that was my human interpretation again. The interior of the tunnels was covered with something that looked like knuckly lichen, skittered over by plenty of small, spindly critters. Things with whorled shells, and things bristling with long hairs. Small for Shroud, which meant topping out at around the size of a human being. We didn't meet anything big enough to have bored the tunnels, but by then I was wondering if they had been carved at all. Instead, the licheny stuff seemed to be slowly eating away at the walls. Maybe where water had started to seep into the cracks, life had followed and just dissolved the rock? I wasn't exobiologist enough to do more than speculate.

'It's quiet,' said Ste Etienne thoughtfully.

Honestly, it should have been me picking up on this, but it had been a while since I'd looked at the local radiosphere. The pod had alerts set, to tell us if signals arrived on the Shrouded's own band, and other than that there was just noise. Except there wasn't, any more. The outside babble of Shroud's life faded away after a few turns, and we were moving through a blessedly simplified electromagnetic environment. Most of the small locals seemed to be finding their way around through scent or touch, and had nothing to say to each other. Only the muted chatter of our escort was left.

In the two big Shrouded nests we'd encountered, the same shielding effect had been present, but there had been so many of the Shrouded constantly talking to one another that the seclusion hadn't been obvious. Here, now we were aware of it, the relative radio quiet was almost creepy.

We took a few more twists. The ball drone was running ahead of us, double-checking the route against its earlier map. We really didn't want to accidentally take the wrong turn and become a permanent fixture of this maze.

It was a strange thing, psychologically. Nothing about our environment should have worked on us. We'd been in the dark since we arrived, and the genuine silence of this place operated on a suite of senses humans didn't even possess. The tunnels beneath this world should not, under any circumstances, have been eerie to a human mind. And yet they were. As though a flicker of the Shrouded's perspective were somehow bleeding into us. We were beneath the mountain, in the underworld. Such places have always spoken to something in the human experience. Mystical, threatening, places of death. Perhaps it didn't matter how alien the world was. All underworlds were one, in our collective subconscious.

I saw it first. The thing ahead of us. Just a shimmer in the lamps initially. Then we were watching it gleam, like a faceted silvery jewel – a dome-backed shell with a fantastical fringe of shining spines. Our lamps refracted from it in a spray of blue-green-yellow, quite the most beautiful sight Shroud had displayed for us. It was clamped to one wall of the tunnel like a mineral deposit of uncountable value. I can't say for sure, but I reckon that's what we both thought, seeing it. We wondered if it was something we could salvage, somehow. Turn up at the anchor point carrying the crown jewels of Shroud to try to buy our way back up the *Garveneer*'s rank ladder. We'd not just be famous survivors, but rich! Obviously neither of us had even begun to speculate what elements the stuff might actually be made of, but the very sight of it said *wealth!*

Until, of course, it moved.

4.3 DARKNESS

The Stranger takes the worst path, of course. I do my best to warn it of the danger, seeing it frustrated in its stumbling journey. *Not the caves*, I tell it, and make sure I draw its notice to the threat. Yet into the caves it goes.

There is a moment when I decide not to follow. Between my segments, I muster enough pieces of memory of when I'd first passed through there. I had been strong and wise when I entered that underworld, and known many things. When I emerged into the face of the wind again, I was weak and scared, and my memory was torn to shreds. I have made the coast my own now, my home amidst the vents and my harvest, but I recall having to build from nothing. It is my eternal curse, that each thing I learn sloughs off me. It falls into the gap between *me* and each Otherlike, so for every step I take towards becoming something greater, I am dragged back. A miasma of ignorance surrounds me, like corrosive air that eats away at what I am whenever I venture too far from home.

I do not want to follow the Stranger into the caves, but following the Stranger is why I sent myself here, and so I do.

At first the calm within the caves is almost welcome. My own thoughts come clear to me, and I feel alert and composed in a way I haven't known since I was within the chambers of home. The small creatures of this place have piping voices and almost no thoughts at all. I let the clarity of quiet aid my deliberations, following the Stranger and its small round child that it sends ahead of it.

I begin to die.

A segment flares a brief siren of alarm and hurt before it is gone. My thoughts rush out to feel across the walls of the cavern, my voice exploring the contours of the cave. I catch a whisper of movement. Spectral feet. But my thoughts dance across the limits of my surroundings and find nothing.

Another segment struggles briefly, reporting barbed skewers in its flesh, a brief burn of disruptive chemicals, then is no more. I rush over again, explore, finding nothing but the dead. I am not many here and panic rises – I cannot think what to do. I am attacked, I am eaten away. Nothing is there. I lash out, feeling only the hardness of the cave.

Unh unh unh says the Stranger, and then again with more strength. Mindless wordless grunting, but I am losing my words with each segment. There is movement all around, but when my thoughts reach out there is nothing there.

Terror rises in me. I give myself to it.

4.4 LIGHT

At first we couldn't work out what was happening. The gleaming thing, with its silver shell, uprooted itself from where it had dug into the cave wall, and we assumed it was going to move out of the way of this circus we were bringing through its world. Instead it attacked. A flurry of motion erupted in the light as it lashed out with a freakishly extending assembly of parts. These grabbed one of our escort and dragged it over. Some sort of mouthpart or stinger or something had gone in. Struggles were brief and then our companion was plainly paralysed or dead. We watched the others swarm around us and waited for them to descend on this opportunistic predator and tear it a new one, like they had with others so many times before.

They went over to that part of the cave, sure enough. But they were pointing all different ways, almost comical as they quested about. The thing was right in the middle of them, hunkered down, eating their friend, and they didn't seem to register it. To us, its shiny shell was so polished it might as well have had its own light show going on.

Then it took another one. Just reached out, almost lazily, and swiped it, right from the centre of them all. By then a second had appeared, and was doing just the same thing. They moved very slowly and carefully, even more so than the regular Shroud gait. Very deliberate, with a weird stop-start motion but still sure in the strike. And our escort just milled around, doing nothing except be killed.

I was signalling them by then. Ste Etienne pulled the pod back to where there weren't any of the silverfish things, and I continued to pelt our escort with three-beat radio pulses, all calibrated with their signature complex inner rhythms.

They came back to us eventually. There was quite a wreckage of prised-open carapaces left behind, but we still drew the majority of them with us. We'd worked out what was going on, by then. The limits of their sensorium.

The silverfish started creeping closer. They probably didn't encounter a feast like this often and were keen to take advantage of it. So we kept sending out our signal like the Pied Piper, drawing our suddenly tractable escort with us, dancing slowly around the encroaching menace. Ahead the lamps caught more gleaming ambushes, and we worked our way around them too. We were slow but, because of their particular predation strategy, they were slower. They were almost completely reliant on nothing seeing their approach.

They never troubled us in the pod. We didn't register as prey to them, nor did the ball drone, although some of the cave's smaller inhabitants bowled it over experimentally a few times. Like the pod, though, it was gyroscopically self-righting, and its meagre signal traffic with us obviously didn't suggest anything delicious was within it.

We took twice as long as necessary to get out of the caves. The silverfish things were dotted all the way through, and we used the drone's mapping to find us alternative routes to skirt around them. They were constantly closing in on us, until the very sluggishness of their approach became a part of their threat. They seemed utterly indefatigable, and if we had ever ended up at a dead end, or had to double back, they'd have torn through our witless escort with impunity.

At one point we became lost. We only realized when we ended up at the edge of a huge hole, a breach into the roof of a far larger cavern below. We almost fell into it, skidding a little down the slope, across the icy organic carpet that covered every part of the caves. Gleams of shiny carapace flickered in the rear lamps. So we could only go forwards, inching around the edge of the sheer drop while the ball drone ran ahead of us to try to find a way back to regions it had mapped.

We never saw what was down there. The lamps couldn't reach so far. I felt my stomach clench at the very thought of it.

Then, soon after that, the drone was reporting open air, the far side of the mountains, a relatively shallow slope and a distinct drop in temperature. We were out, and we were crossing the notional equator, heading into the outer hemisphere of Shroud.

For a moment this was a cause for celebration. Then Ste Etienne remarked it meant we only had a whole half-world left to cross. The entire distance from this midpoint to the outward-facing pole where the anchor was. A far greater distance than the trek we'd already covered.

Neither of us drew the other's attention to the various telltales signalling the condition of the pod's systems, especially the leg joints. The ambers and oranges, and the occasional red. Happy, festive colours all. Autumnal, speaking of a bountiful harvest of mechanical issues we'd be reaping soon enough. Neither of us said that it simply wasn't going to work. We just parcelled out the maintenance tasks and pretended the limited amount of fix-it work we could manage, from inside the pod, would stretch out the operational lifespan of our vehicle for just long enough.

The Shrouded, our diminished escort, still seemed subdued. They clustered close to us while we made our repairs. Nothing shiny and nasty came out of the cave after them, and maybe whatever sensory trick the silverfish had used didn't do so well in a world awash with radio signals. Maybe at that point they registered as a kind of negative space or something. We couldn't know, but I reckon we and the Shrouded were sharing a truly interplanetary relief at having left the things behind.

We had no idea, of course, what was waiting ahead of us.

We crossed the equator, descending the mountain slope. The temperature was a solid ten degrees colder than it had been on the far side, here in the wind shadow of the mountain. Because the wind was warmth, or Shroud's pitiful nearest equivalent. It carried the feeble heat from the half of the moon constantly faced towards the brooding radiation of Prospector, the planet. In the regions we had travelled through, the heavily adulterated composition of the local water table had resulted in, let's say, a fluid relationship with freezing points. That and the interference of the Shroud ecosystem. What should have been a solid icy world had become something more complex and amenable to life. Here, it was different. Winter had come to Shroud, or rather we'd come to Shroud's eternal winter.

When our lamps picked it out, we didn't know what we were looking at. We ran diagnostics and physically cleaned the cameras, because it seemed like an artefact – like visual static glinting out of the gloom as we descended towards it. Then we came closer, and it was still baffling. We moved right up to it, studying the world ahead of us, which was a tortuous honeycomb of yellow-white feathers, extending as far as our light could carry.

'Ice,' said Ste Etienne eventually. But ice didn't behave like this. You didn't have ice forming a vast three-dimensional sculpture, just suspended on nothing-at-all, reaching further than we could see in every direction, above, below, ahead. A chaotic interwoven mass. Some of it was delicate as candy-floss, other parts showing solid, twisting pillars a metre thick, with commensurately large paths leading into the interior. As delicate and intricate as the fine structure of a snowflake, but magnified beyond reason.

In the end we took a sample. Not for the chemistry so much as the structure. We carved a segment out of a thirty-centimetre-thick strand and found a bundle of hair-thin filaments in the middle of it, which appeared to be living. Or previously living, before we'd amputated them at either end. There was some infinitely fine network of life here, and the ice had formed as armour around it. The ice was a forest, a living biome.

We forged inwards, amongst it. The Shrouded weaved their way through the glittering sculpture-scape with minimal contact. The pod, lump that it was, just smashed a pod-shaped path through everything, leaving a wreck of broken ice in our wake.

I ran some models as we trekked through it. The dense atmosphere of Shroud we had slogged through thus far had been further obscured by a haze of part-frozen water-ammonia-methane mix, which had been adding to our maintenance woes during our trek. What was surrounding us here was just the same, plus the drop in temperature. It had iced out of the air in these fantastical, delicate structures, along the thin tendrils of whatever the hell kind of life it was we'd just excavated. Which, in turn, didn't need to build any kind of structural exoskeleton to keep itself in shape, since the ice had done the work for it. Or maybe it ate stuff caught in the ice, or there was some other interaction entirely going on. It was just another mysterious Shroud thing, and by then we were becoming almost blasé about them.

We discussed whether this was going to be the dominant terrain from now on, between here and the pole. Our uninformed consensus was probably not. We knew the pole itself wasn't covered with this kind of freeform ice sculpture, because we'd seen what was there when installing the anchor. Just ice and rock; an ecosystem almost spectacularly barren compared to anything else we'd seen. The cycling wind cells didn't reach as far as the pole itself. The various climatic engines that turned the crank of the Shroud biosphere didn't bring either heat or biomass there, or enrich the air. It was a stagnant little backwater, coincidentally perfect for setting up the root of our space elevator. Also the coldest point on the planet, so any moisture in the atmosphere would have frozen out long before reaching there. Hence, none of the complicated nonsense we were currently navigating. Ste Etienne reckoned there were warmer regions ahead too, blessed by the circling winds, where we'd find less freezing and probably more life.

That was our mistake: to assume this ice-feather biome was short on life.

Unlike with the silverfish, our escort plainly registered the presence of the locals. Their signalling picked up, and their movements became more agitated. From their point of view, they were probably trumpeting the alarm, shouting right in our ears about it. But we didn't understand, of course. To us they were just bustling a bit more and that could have meant anything.

And we didn't even see the threat initially. It struck from beyond the reach of our light. One of our escort had stopped and was flaring out various parts of itself towards the darkness, a threat display, making itself bigger for the radar and sonar of its enemies. Then it exploded. Literally just tore apart in a glittering flash. Shrapnel from its sundered casing rocked the pod and smashed one of the arm joints. Ste Etienne swore and instinctively stopped walking, hunkering down. Then the assault began in earnest.

Our escort scattered, reformed, accelerated. Entering that hyper-energized state where their movements were abruptly five times as swift, like a speeded-up film. They made sallies out of our sight, then returned, always fewer. We caught the flash of missiles passing through the limited circuit of our visible world, leaving a glittering trail behind them in the air. Here we were, thinking we'd encountered the most Shroud had to offer in the way of intelligence, and just over the mountains was something that had invented guns.

Our Shrouded escort next chose a direction and struck off in it determinedly. Ste Etienne threw us after them, at the top speed the pod could manage. Their communications band had gone wild. Around us, the baroque, ethereal ice sculpting wheeled and shone, translucent and opaque in turns, forming a multi-layered puzzle for the eye. I had no idea how clearly it came to the senses of the Shrouded, whether it was a solid maze or like draperies of gossamer.

We finally saw one of our attackers. It was a globular thing hunched within a minimalist frame of sticks, like a frog with a mobility aid. Its skeleton was just a scaffold that was mostly open space. We studied it later in a single frozen frame, where the precise lever arrangement of its alternating limbs was revealed far more clearly than we'd ever seen before. The way everything hinged around the rigid sections of the outer skeleton, and the system of every limb section feeding nested levers into the interior, from where the resident could puppet the whole frame about. This wasn't what we were fixated on right then, of course, because we were seeing Shrouded missile weapons in action.

It bloated out until it was filling its openwork cage, bulging from the broad gaps. As though it was a cartoon character who'd swallowed a grenade. A moment later it *spat* a bolus of something right at, and into, the closest member of our escort. There was a bright white flash and the luckless Shrouded was ripped apart. In the heart of it, tearing through frame and living material both, was a spiky ball of white.

Ice, I realized. The deadly weapons were nothing more than water. Water which, almost uniquely, expanded when it froze. And it shouldn't have frozen that explosively, but there were ways. Hydrogen peroxide and a catalyst within that bolus mixed together on impact – I modelled it later. Maybe I was way off the mark but those spitters had certainly got the formula down. They were all around us by then. The fleeting lances of their missiles zipped through our wheeling, juddering arc of light like angry wasps.

We were all fleeing now, us and the Shrouded. We kept

catching glimpses of more spitters, mostly clambering above us as they picked out the spars and flourishes of ice which would bear their weight. Unlike the silverfish in the caves, they didn't seem to be interested in eating their kills. Perhaps they knew our Shrouded of old – no reason to believe our new friends were particularly good neighbours. Maybe we had walked right over their nest or nursery, or smashed up their home too much on our way through. Or maybe it was some other alien thing a poor human could never understand.

Something hit us, hard enough to send the whole pod staggering. A second impact from the other side actually stabilized us again. Then we were limping, and I saw one of the legs go all over red indications, and then black. Ste Etienne made a high, incredulous sound.

'What happened to the diagnostics?' I yelled at her.

'Diagnostics are fine!' she yelled right back. 'Leg's gone.'

'What?'

'Just . . . fucking . . . gone,' she gulped out, voice shaking. We were limping along now, rolling drunkenly like someone who's been in freefall for a year and suddenly has to cope with standard G. The pod could, apparently, manage with three out of four legs but I knew damn well it definitely wouldn't with two.

I decided then that, deadly water pistols notwithstanding, these things couldn't be that bright. Because they absolutely had us if they wanted us, but they weren't targeting the pod's limbs, just letting fly at us randomly, and possibly wondering why we didn't explode. They had evolved in a world of basketwork monsters. Environmental survivability meant we were far more enclosed than any prey or rival it was used to driving off.

They were merciless, nonetheless, and very good shots. Ahead of us, in the vertiginously wheeling light of our lamps, we saw our escort being picked off, with every hit a fatality as the fluid flash-froze into spectacular, gut-ripping sculptures.

Then we were skidding down-slope, rolling over occasionally before the gyroscopes tried to right us again. Jolted and yanked against our straps by the heavy hands of Shroud's gravity. The lights died, flared on too bright, then died again. I had a brief glimpse on the screens of one Shrouded – the last survivor of our entire escort? – skittering madly to one side to avoid being crushed by the pod's wild tumble.

And then we were still. Or no, not still, faintly swaying, on some moving surface, sluggishly returning to a level orientation, in the pitch dark. No lamps shone outside, no lights inside, no cameras. There weren't even any diagnostic lights for a good twenty seconds, before some kind of autorepair was kind enough to give us a bank of red indicators. The impacts on the hull had stopped; our mad tumble had taken us out of our attackers' territory or beyond their reach.

'Mai?' I whispered. I couldn't make my voice go louder. I could feel tight bands of pain across my chest, gut, thighs and calves, everywhere the straps held me. They'd saved me for sure, though. If I'd been thrown clear of the couch then the battering of the fall would have turned me to jelly. But I was hurting and, right then, there was nothing to shoot me up and numb the pain.

'Mai?' I said, more urgently, because I'd just clocked what would be considerably worse than all my current woes and that would be the addition of a corpse.

There was a long, hissing exhalation in the dark. I could have wept.

'Going to need you,' Ste Etienne said slowly, 'to fix something for me.'

'Lights?'

'Sure, probably going to need lights. But then the pharma dispenser online, please. Think my arm's broke. Ribs maybe. Hurts like fuck.'

By then the diagnostic system had followed its priorities by making sure the diagnostic system was properly functioning, after which it metaphorically turned to me with spread hands and a *what-can-you-do?* expression. For the next thirty minutes, acutely aware of the pain Ste Etienne must be in, I squinted sideways at the dim glow of my screen and battled my way through menus, until I had re-routed enough power and logic connections to get the pharma units working again. Another ten minutes saw sufficient painkillers and stimulants printed up for both of us to be feeling artificially a great deal better. If not actually about the situation.

We still had three legs on the pod, although one of them had a seized knee that took some fixing, and which didn't look as though it'd hold for long. We told each other how lucky we were, and neither of us mentioned that we still had approximately a quarter of Shroud's entire circumference to travel before we made the anchor. As though we might run into a friendly service point just a little further down the road, where we could get all these minor mechanical issues ironed out.

Ste Etienne let out another long, slow, breath, as though she'd been holding it for the past hour.

'Got some of the lamps ready to go, and some of the cameras. About half of each and not exactly the right half either. Want to see what's out there now?'

'I am picturing,' I said, 'a whole load of monsters with cutlery and napkins. And arms ending in God's own tin-opener.'

She gave a truncated laugh which she plainly regretted.

We turned the lamps on.

4.5 DARKNESS

No
thought.
Just
run.

Threat.
Flee.
Lost.
Wait.
Hide.

Call.
Quiet.

Call.
Quiet.

Run.
Run.
Run.
Lost.
Scared.
Threat.

Run.
Wait.

Call.
Quiet.

Wait.

Hunt.
Feed.
Wait.

Call.
Quiet.

Lost.

So
lost.

Lost.

Walk.
Walk.
Call.
Quiet.
Scared.

Mind
blank.

No
thoughts.

Walk.
Lost.
Walk.

Call.
Response.

4.6 LIGHT

The pod had not come to rest on solid ground. There was no solid ground for it to rest on. We were cupped, almost poetically, in a bed of flowers. Well, not flowers, obviously, and what they turned their hungry faces towards was not the unseen sun. *Deep sea creatures*, said my imagination. Crinoid things. With feathery hands reaching ragged fingers up and out, their gill-like fronds trailing in the breeze. They jostled together, barely a hand's breadth of space between any of them, groping upwards to where a swirl of snow blew past. Ice fragments, perhaps. Organic detritus. I thought about all the life carried into these colder climes by the circling wind cells. Probably a lot of it precipitated out when it hit the temperature gradient. This whole ecosystem could be fuelled by opportunistic scavenging.

'What's below us?' I asked.

'That's the real question, isn't it?' Ste Etienne agreed. Right now we were sitting across four or so of the flowers, putting quite a dent into the whole patchwork quilt of the landscape. Plainly there was some kind of support structure below, but these things were reaching *up*, and we'd already seen how Shroud life had a tendency to build its own living landscapes. It could be sea down there, or ice. Or it could be spinning wheels of teeth and industrial grinders for all we knew. From what we'd seen of the world, that felt entirely possible, and the topographical data from the blimp drone hadn't penetrated this living outer layer.

'How are the legs?' I could see the sanguine-slanted kaleidoscope of the indicators, but I needed to hear something reassuring from a human voice.

'Give me level ground and I can make this thing walk, at least for a while. On this?'

Something passed overhead. A suggestion of vast pale vanes and finger-like struts. I whimpered, even though anything light enough to be carried about on the wind was the last sort of creature to pose a threat to us.

'I think,' I said carefully, 'that I am reaching some kind of shock limit.'

'Uhuh,' grunted Ste Etienne.

'I don't know the medical term for it,' I said, still picking my words as though I was walking through a linguistic minefield. 'But a lot has happened in the past few days and I'm – having difficulties.' Breathing, for example, was suddenly becoming surprisingly hard, until the couch shot me up with something new, which interposed a numbing blanket between my panicking mind and my labouring lungs. 'Mai. It's too much. I know we're supposed to be able to carry on no matter what. I know that's the Concern way. The worker's way. We do our part. Did you sing that song, in school? "In the dark and in the cold, under fire and—"'

'I fucking hate that song,' Ste Etienne spat.

'Mai! It's *the* song. The Worker's Hymn, right? "We earn our wage-worth with our toil, so that humanity can—"'

'I bet you sang it loudest in the class,' she said disgustedly. 'I bet you were given the Cobalt Star of Praise.'

'The what?' I said, thrown.

'The Cobalt Star of . . . You know, when someone's really lived up to Concern ideals, been the model little child-worker, sung the songs, drawn the pictures, done everything *right*. Won the games, got the highest score, passed all the learning aptitudes. The Cobalt Star of . . . You didn't have that?'

'I've never heard of it in my life,' I said. 'I mean, I'd have been all over it if I had. Would have been over-achiever of the week, every week, if anybody had cared back then.' It was ridiculous, talking about this stuff. I couldn't imagine anything less relevant. Except it wasn't about Shroud, or the pod, or the fact we were very definitely going to die, unmourned and unknown, on this alien world. It was about something that couldn't possibly matter, and sometimes that's what you need.

'So, go on,' I prompted.

'Go on what?'

'How many times did you win the Cobalt Star of Praise?'

Ste Etienne muttered something.

'Speak up for us at the back,' I enjoined her.

'I mean, apparently it was just some made-up thing that our tank's teaching system had, so I don't see it really matters,' she snarled.

'Was that five times or six?'

'It was once,' she said flatly. 'And they'd taken it back off me before the end of the learning cycle 'cos I lamped one of the other kids. Happy?'

I was not happy. I couldn't even draw a map from where I was to happy, not in the dark like I was. But it helped. I felt around the edges of my equilibrium. Still fragile. Still pieced together and held in place with spit and tape. But not about to come apart any more, into all those razor-edged fragments.

'Let the ball drone out,' I said. 'If it still works. See what's down there.'

'It can't walk, but its sensor suite's still going,' she agreed. 'Let's see.'

It was solid down there. We took ourselves tentatively down after the drone, half climbing, half abseiling, hooked to a dozen of the crinoid things, and we discovered it was ice. Actual ice. We'd reached a temperature where, no matter the impurities, the sea of Shroud had frozen. The stalks of the detritovore flowers above us projected through, each with a clear shaft around it where some chemical or thermal process prevented the ice from forming. The surface of the ice was riven with fractures and impurities, but rather than being a shifting jigsaw only held together by some benthic horror, it was solid enough to bear the pod's weight with barely a complaint. And flat. Just about the flattest real terrain we'd seen on this whole benighted world. We consulted the data from the aerial drone. We had to make a best guess at where we were, but we had a decent ballpark idea of it. The easy going wouldn't last, though, according to our long-range forecast. Precious little of Shroud was actually flat, even on the sea.

On our two and a half legs, we set off again. Even though there wasn't much point, there was even less in staying still.

And then, later, we encountered the pylon.

I don't know what else to call it. We initially thought it was just one more Shroud native. A great metal tower sticking up from the ice, crusted over with frost and barnacle-y life. The exposed iron hadn't rusted but was wind-abraded enough that it must have been there a hell of a long time. It was constructed in a similar way to the outer structures of Shrouded life we'd seen, especially our vanished escort, of whom there was no sign since the last violent ambush. It just went up, though, and it was hollow. No living thing inside it. The abandoned husk of some deep-rooted titan, perhaps? We set down beside it for some jury-rigging and fixing, and kept scanning the lamps and cameras up and down it. There was a regularity to it that wasn't like the organic engineering of the local caddis-case skeletons. A lack of tooth and whorl. Something that spoke far more of the *made* than the *alive* to the human eye. I ran a brief analysis of its stability, which returned high marks on 'efficient use of resources for maximum strength'. Which meant nothing, because natural processes can hit on ideal solutions by evolutionary trial and error, but still . . . If we'd wanted to build it, we'd have done it something like this. Because the basic maths of weight and stress don't really change from world to world, once you account for local variables like gravity and atmospheric density. I thought of those learning sessions in the habitat tanks. It would be the sort of exercise they'd set: design a tower with the greatest possible strength-to-materials ratio. Winner gets a Cobalt Star of Praise.

I didn't say anything foolish to Ste Etienne and she did me the same courtesy. Besides, the thing was plainly very old. It wasn't like we were about to encounter the Great Pylon Civilization of Shroud. It was just a weird alien thing on a weird alien world, and we had no context with which to interpret it.

The going got tougher from there on. The ice sloped upwards. We thought it was land at first, but the pod had a geophysics suite which confirmed nothing underneath us had any 'geo' to it. So the ice had been thrust up by some past convulsion, perhaps, then smoothed over by the wind. By then the profusion of detritovore stalks had been left behind. Instead, the ice itself was a biome. We saw large patches of discoloration – things living within the ice itself and maybe contributing to its irregular contours. Some were just stains, like subsurface lichens. Others moved: worms and blobs and films, creeping about through fissures and fractures, larger than a human being, and plainly visible to us when we turned our lamps and cameras onto the gleaming surface. As though we were seeing a living body with pulsing organs and vessels exposed. There were bubbles in the surface sometimes too. We only found this out when we cracked through the first of them, shattering the ice above it and almost toppling in. A hollow world, four or five metres across, was lined with complex, gnarled life and a haze of little hand-sized scuttlers. I wondered if it would all die now we'd broken it open. If we'd just exterminated a completely self-sufficient biome with our human blundering.

After that we kept the geophysics suite pointed forwards to pick up any more unexpected cavities. Not because we were fanatical alien conservationists, but because if we'd actually dropped into that hole then the pod could have cracked open. Ste Etienne had over-engineered it for shock tolerance, given the heavy gravity, but we hadn't exactly been treating it well. There were probably all manner of weak seams around its periphery by now.

They found us soon after that. Once we'd started labouring up the slope. A slope we'd known was there, and that the aerial data reckoned was entirely navigable, but it was different with three legs. We were constantly having to reposition the feet so the pod didn't end up trying to lean on the missing fourth. Ste Etienne tried to make the walking algorithm accept the amputation but something in the routines kept resetting its assumptions. We were having to do more and more of the walking by hand, as our automatic systems failed us.

Halfway up the slope, we were attacked. At range, again. Not with water this time, but something that was solid when it hit us, then sprang out in a tangle of rods and pieces, scrabbling at our hull before sliding off it. We thought a living thing had jumped at us, honestly. When we turned our cameras on the detritus, though, there was a weird spiky tangle of jointed rods and hooks. Something that might have made a real snarl of an open-plan segmented exoskeleton.

The next shot made a serious mess of our damaged leg. Aimed at it, too. The snipers had watched us limp, identified the weak joint, then sent a bolas right into the damaged knee with pinpoint accuracy. We rocked and ground about on the suddenly immobile leg. Something groaned, deep and pained, communicating to us through the fabric of the hull. We sagged, tilted sideways and then adjusted back to upright. When we tried to step again, the best we could manage was a halting circle around the stilled limb. A muted alarm started making itself known – a single voice that knows it won't need to wait long for the rest of the choir.

For a long moment no further missiles came our way, and neither of us said anything.

A foot came down, within our view. A piece of mechanical engineering, really, rather than a living thing. Except on Shroud mechanical engineering had arisen out of evolution.

As though, on Earth, a crab or a snail or some other creature had been gifted with the ability to improve the design of its carapace, its shell. The clever behaviours that instinct could throw up, like the farming activity of ants, had been applied to a weird bioengineering here – self-improvement at a species level. We had a good look at that foot. It had been planted very carefully out of the furthest extent of the pod's reach but well inside our light, because its owner didn't know what light was. They'd clearly watched us flail and pivot, and worked out exactly where the limit of our grasping range was. Every predator lives in fear of crippling injury, after all. Even, it seemed, the ones which could build their own legs.

This foot had cleats built into it, for the ice. It was a marvel of articulation. It belonged to something very big, because the leg extended out of our sight and the body was lost to the darkness above.

Ste Etienne growled, deep in her throat, and raised one of the pod's arms like a scorpion's tail, or a striking snake. A threat straight out of Earth's playbook. But apparently something communicated across the boundary of space and species, because the next missile clipped that arm off entirely.

It was the sheer clinical casualness of it that chilled me. No snarling, slavering monster scrabbling at our hull with teeth and claws, but some long-legged, dispassionate thing with a slingshot, carefully identifying which part of this odd ball might be dangerous, then neutralizing it.

It loured down into the reach of our lamps, a ghostly lattice of struts and whorled segments. The leg we could see was its forward limb, on the right. Set back from it to the left, the next limb spidered off into darkness. Inside the body, we saw the living heart of it, the actual Shroud organism.

No – organism*s*. There were multiple caterpillar monsters, with springy pelts of gleaming metal hair and branching tendrils reaching out left, right, left, right, down the lengths of their bodies, grasping the levers of power.

In the belly of the body's hollow frame we saw something ratchet-taut. I'd thought *slingshot* before, and now I saw the thing, that was just about exactly what it was. A little crossbow of flexible arms and some kind of elastic tendon, with a carefully folded bundle of sticks poised within it, ready to fire. Given the gravity and the dense atmosphere, I didn't reckon they had a great range, but if you had the only crossbow on the planet then that probably didn't matter much.

We didn't realize there was another one coming in behind until it pinned us down so that all our leg joint indicators suddenly shifted another shade redder. The alien foot was clamping one of our remaining arms to the hull. Again, they had worked out exactly how the pod was put together and how to take advantage of that. We'd seen a lot of examples of how Shroud life might or might not approach some kind of recognizable sapience, but this was by far the least welcome.

The second creature was another can of worms in a single frame, the whole nest of them working together to make their jointly operated property walk around and attack hapless offworld explorers. Which is exactly what it did. Three long arms flexed out from where they'd been folded inside its carapace, then ripped our leg off. One of the formerly better-functioning legs. They just braced and flexed, we felt a pathetically small shudder, and the leg was gone, all the readouts dark. We were now down to two and, short of a little dragging, that formally ended our excursion on

Shroud. Not, I should stress, that we were ever really going to reach the elevator anchor point. But it had been good to have even a fantastically misplaced hope of our chances.

'Mai,' I said.

'Juna,' she told me. 'Don't.'

'It's not your fault.'

'Damn right it's not my fault!' she shouted at me, doubly loud in the pod's close confines. But then, quieter: 'It's my fault. I should never have built these fucking things for them. It was always a stupid idea.'

'I mean, we'd have died in orbit if you hadn't. Like Skien and Rastomaier.' Apparently I was going to be philosophical right now.

'Better that,' Ste Etienne decided, 'than this.' We'd had quite the little suite of damage alerts wittering at us, after the immobilized leg. As the two – no, three now – looming monsters bent closer to us, she turned them all off, one by one. If you're going to die, you don't need all that ruckus. And we were, I could finally admit to myself, about to die.

'You want to record any final message? In case someone finds the wreckage? I mean, we've got a black box, and maybe it'll survive whatever they're going to do to us. So if there's anyone you want to cuss out or pledge your eternal devotion to or something, now's the time. Sing the fucking song if you want to.' She was working so hard at being the tough one, but by the end of her little speech her voice was a juddering mess.

I opened my mouth to say something. One of the alarms was still badgering us. I went to turn it off, just for optimal making-peace-with-your-inevitable-end conditions, then saw what it was.

'Oh God,' I said. 'It's them.'

'What's what?'

The EM band the Shrouded had always used, from the first encounter in the briar forest, through the sacrificial chamber where Bartokh had died, to the vertical ant farm and all our several escorts – it was alive with chatter. This wasn't a maintenance alert, Ste Etienne having turned them all off. This was the alarm I'd set to let us know when they were signalling.

I sent over the password, our three-beat signal, hoping against hope. For a moment they went still.

But these were not the same individuals, colony, community, whatever. Our escort had been wiped out within the ice forest. These Shrouded in their multi-monster vessels were complete strangers to us, and they didn't understand.

Bending low, with the dispassion of surgeons, they began to take us apart.

4.7 DARKNESS

I am quite engaged in exploring the construction of this thing, when an odd thought strikes me. Strikes me from the object, in fact. A simple signal, poorly attempting to mimic genuine cognition. Which, of course, only whets my curiosity more. I've never come across any living thing that builds itself like this, but whatever dwells within it is plainly relatively complex. When I have extracted it from its exterior, I may provide it with materials and watch to see how it constructs such a solid and impermeable carapace. It will be, I predict, a purely intellectual exercise. Whilst the creature's shell is certainly defensible, it is a direct impediment to mobility. With only a modicum of carefully applied violence, I have rendered the creature entirely immobile.

It thinks its simple thoughts to me again. My own thoughts run along its curved exterior, reporting back to me on holes and seams. I have my segments pinion it between them, holding it down, digging their hooks into anywhere that seems promising. Another consignment of me from within my home switches tasks to help analyse the curious – unique – construction. I am going to learn something very interesting today.

I begin to apply force. My first few ventures find no give in it, but then I locate where one of its limbs was, before I removed it. There are uneven pieces in that area, and gaps that plainly give access to deeper layers of the creature's shell, and probably the entity itself. The limb must be moved somehow, after all. I fit my hooks from multiple directions, calculating the ideal angles with which to obtain the maximum force, while expending minimum effort.

I pull, and parts of it give way. Stubbornly at first, and then more readily as various points of attachment spring apart. Very good. And, in itself, fascinating. I cannot see how or why anything would construct so excessive and weighty a shell, but the methods involved are entirely novel.

It is unfortunate that, sometimes, the only way to find out how something fits together is to take it apart.

I brace for the next effort, already feeling the welcome weakness in the thing's seams. This exercise promises to become easier as it progresses. I shall soon have this creature entirely unshelled, and then I'll see what is to be learned from it.

A moment of hesitation arises in me.

A memory. Unfamiliar. All the context scoured away. A fugitive thought. *Don't.*

The helpless, thick-shelled creature's thoughts leak out, a mad babble of weeping and mewling in many voices, but in the midst of it is that simple rhythm. A basal, bestial grunting, the imitation of true thought.

A memory. A past association. I have met this before.

Not *I* I, but some fugitive part of me. Segments are always coming and going. I am the master of a great expanse of the world here. I have many neighbours, and we inherit thoughts from one another sometimes.

I have inherited thoughts. Recently. Broken, incoherent thoughts. Pieces of memory I could do nothing with.

I hold the beast between my segments, wanting but a little flexion to open it up, sensing the secrets within it pull taut in my eager hooks.

Unh unh unh.

Another feeling then rises within me, that I should respond in kind. Why? I no longer know. The context for this urge has been carved away by trauma and loss. Only it is left.

I sense a great void of ignorance that I am brushing the edge of. True ignorance. Not just *What is beyond the furthest landmark that my neighbours know* ignorance. A gap of knowledge bordering on the existential. It is a long time since I have felt that chasm yawn. I have been unchallenged master of my home for an age, growing large and thoughtful, with my embassies and splinter-selves ranging far and wide. I have been here, in fact, as long as I can remember, and that is a very long time indeed. If I assemble all my oldest thoughts I can even recall a far distant past when I felt I understood everything in the world. But it all fell away, and I was left with an appalling ignorance where infinite understanding had been. A moment of grace I have been trying to reclaim for all my lived experience.

I relax the tension of my hooks, just a little.

Unh unh unh, thinks the creature, and this time I reply.

4.8 LIGHT

They stopped. For a moment we just stared, each into our own personal abyss, and waited for the next terrible thing to happen.

But there it was, that signal. The single stick-thin bridge between Earth and Shroud. Resurrected, somehow.

Ste Etienne let out a shuddering breath.

'Our guys,' she said, into the relative silence. 'The ones who came through the cave and the ice spitters . . . They must have . . . I don't know.'

Must have found some other community of their species, and somehow communicated . . . what? Not enough to have them roll out the welcome mat when we first entered their territory, certainly. That hadn't been merely over-enthusiastic rough-housing we'd experienced. They'd been trying to kill us. Yet now they weren't.

Through the functioning cameras, the Shrouded hunched over us in their outsized communal bodies. Vehicles. Exoskeletons. None of the human words really described what they were. These structures that most of the mobile life we'd seen here built, so must be something evolved and instinctive. Except the Shrouded, *our* Shrouded, they of the shared signal, could plainly build an outfit for every occasion. Whilst that might just about fit into the box of 'instinctive',

I'd revised my judgement on that when we met the amphibious detachment of them before. And here, out on the ice, they'd built giant adapted striders, with crews and artillery.

Those three beats came again, and again, and I made sure we replied in kind every time. I'd given up trying to be clever with Fibonacci sequences and other parlour tricks. If we had one in a cage, under lab conditions, then maybe we could have got somewhere, but right now things were too chaotic for the Universal Language of Mathematics to really achieve much purchase on the situation.

'I mean,' Ste Etienne said, after another long breath, 'they've still messed us up something proper.'

We only had two legs left, both on one side of the ring. And two arms, now free to move because that pinioning foot had been retracted.

'Can we . . .' I looked about, switching from camera to camera. 'Where's the leg. The one we just lost. Can we reattach it?'

'Maybe. Wait . . . Shit.' Ste Etienne found the leg, sure enough. It was lying on the ice towards the edge of our circle of light. In pieces. That was the clincher, really. In pieces not because it had been torn apart in a frenzy of destruction, but because it had been carefully dismantled into twelve or so separate sections. One of the Shrouded had been puzzling over it – that was the only way I could interpret what we were seeing.

'Well, this hasn't made our lives any easier,' Ste Etienne complained. 'Fine. Let's at it anyway.'

We had to drag ourselves over. We circled the two legs around until they could push us along, and used the arms to stop us rolling forwards. It was a painful, wretched progress over to the neatly arrayed pieces of leg. Then we had to reposition ourselves so the arms could reach both the pieces and the stump. It was going to take a while, but we had all the time in the world, until some other vital component gave out, and then we'd have no more time. As always, what were our options, precisely?

We were only two pieces into the painstaking process before the Shrouded lost patience with us and descended again, lurching into movement at the same exact instant. Doubtless there had been some lively chatter on their channel about how best to torment the aliens. The pod was shouldered away, with the remaining pieces of leg plucked from our grasp and sight. Ste Etienne was swearing in a steady stream of profanity, while I signalled and signalled, three beats, over and over. Sometimes they replied and sometimes they didn't.

They then began spinning the pod on one axis or another. When we'd had four legs on a rotating ring, the gyroscopes made sure the actual sphere of the pod maintained a proper up and down for us. With the Shrouded yanking us around and turning the entire pod on its head, that luxury was denied us. We jolted about in our couches, one moment pressed into the gel, the next hanging painfully from our straps, which cinched into flesh already bruised from our last tumble. Each time we ended up upside down I couldn't breathe, the whole exaggerated weight of my body pressing against the restraints with a force that my feeble muscles couldn't overcome. My ears were full of the awkward sounds that the rough treatment wrenched from me, gasping, sobbing and squawking. When they finally gave up on rolling us around like a marble, and our world slowly righted itself, I just lay there and wheezed, waiting for the dancing spots to clear from my vision.

'Mai?' I croaked.

She made a sound. I remembered that her arm was broken. While her system would be swimming with pain meds, she'd probably suffered what we'd been through more than I had.

'Speak to me,' I said. 'Please.'

'Just. Wait.' Spoken through clenched teeth, surely. I managed to access her vital signs. She was losing blood, but even as I watched this tapered off. I read a screed of dispassionate notes about compound fracture trauma and shuddered.

'Fine,' she said. 'I'm here. I'm fine.' Sounding a little slurred as the stimulants and the painkillers duelled it out in her system. 'What're they doing?'

'They're . . . um.'

They were observing us. I was an old hand now. I could decipher the activity which meant their communication and the EM buzz of their sensory suite was bouncing signals off us to keep track of where we were and what we were doing. We weren't doing anything, obviously, because of the leg situation. Except that situation had now changed.

They had put the leg back on. Having reassembled it, they'd then reattached it, a precise and careful reversal of all the damage they'd done. Except it wasn't, really. The physical sections of the pod's leg were reattached, but nothing was linked up. The hydraulics and electrical connections were still a mess of severed ends. I imagined some Concern instructor cussing out the Shrouded as bad students. *Call this a repair?* But it was a repair to the extent of their understanding of mechanical advantage and leverage.

The rest we could do ourselves, anyway. Ste Etienne could

probably have done it in her sleep, honestly, given she designed the thing. But the diagnostic system was showing me all the broken connections, set out clearly with exploded diagrams in the same format as every Concern technical readout. A format I'd learned when I was a kid in the habitat tanks. Because when you're training up a mass workforce to send anywhere and everywhere, then having universal tech notation is a really good idea.

I worked through it with our two functioning arms, using trial and error, re-establishing connections one by one. The pod had been designed for rough use, after all. Whilst nobody had foreseen all the crap we'd been through, a certain amount of on-the-fly repairs were par for the course. Every component was robust, and a lot of it would connect itself when the ends were brought close enough. It was all absurdly similar to the construction toys they'd given us as children. All those brightly coloured pieces you could fit together any way you liked, to make little toys that walked around or pinged spare components at your classmates.

The Shrouded watched. Not *watched*, but stared eyelessly at every move I made. Our previous escorts had never been entirely still, but this mob, in their big suits, were utterly fixated on everything I was doing. *Learning*. I couldn't think of it any other way. They were observing and taking note of how to fix human vehicles, as though they were going to be auditioning for our jobs.

'Not bad. Make an engineer's mate of you yet.' Ste Etienne still sounded very weak. I put the leg through its paces. There were three arcs of freedom where the joints didn't work properly, or weren't responding to commands. But when I positioned the leg under us to lever the pod off the ground, it bore our weight. For how long, who knew? But right now it was holding and we had three semi-functioning legs again.

'Are we all friends now, you reckon?' I asked her. The Shrouded were standing over us still, as the pod staggered about in their midst.

'Sure, sure, best pals,' Ste Etienne said, sounding preoccupied. She was accessing the aerial data. Working out where we were and which way we were facing.

The words were crammed right up in my throat and I didn't want to say them. We had three legs, one of which was the worst kind of jury-rigged rubbish. The other two were also showing a variety of damage indicators that would, in any less isolated scenario, have seen them replaced long before this, not fit for purpose. The actual hull of the pod was displaying a fascinating variety of stress fractures and weak seams, both from the mauling the Shrouded had just given us and from every other punishing moment of the moon's hospitality that we'd gone through up until now. It wasn't that Ste Etienne had flunked the design or the building. The pods had never been intended for this kind of protracted and self-sufficient use. Our chariot had already exceeded its intended parameters by several orders of magnitude, and Ste Etienne deserved some kind of commendation, honestly. But it wasn't enough to keep us alive and cover the distances between here and the pole. It never had been.

To put that bleak and inarguable truth into words would have been a betrayal, somehow. We'd neither of us explicitly confronted it before, so how would raising it now help, exactly? An admission that we might as well have banjaxed the safeties and loaded the internal atmosphere with carbon dioxide or something. Gone quietly into that final night. All we had striven for, everything we'd endured, the pain and the anguish, it could only ever have been for nothing.

I heard a story once. It'd been one of a set about how terrible things had been on Earth, with the intent of showing that even life elbow to elbow in the habitat tanks was better, and how generous the Concern was to look after us like this. According to this tale, way back in the Early Industrial period on Earth, there was a disaster in a mine, and they just sealed up the shaft afterwards, because everyone was obviously dead. Only later, when they opened it up, they found the bodies of the miners just below the surface. They'd basically dug all the way up from the very bowels of Mother Earth. All the way up the flooded, collapsed shaft, working together, desperate but determined. Until they reached that final barrier – the man-made blockade that had entombed them. And then they died. Their herculean effort had brought them almost, but not quite, far enough.

We weren't even going to get as far as *almost*. I knew it, and Ste Etienne knew it. And I wanted to say it right then. I wanted to scream it. I wanted to point to the fact that our atmosphere recycler was faltering now, and the temperature within the pod was three degrees warmer than it should be, and climbing, and at some point it would probably plunge catastrophically the other way. I wanted to go into bitter rhapsodies about the fact the pharmacopoeia was running out of stim, sedative and painkiller variants that we weren't developing a tolerance for. I wanted to scream, basically. Every time I opened my mouth to say something calm, practical and task-focused, a vast vomitous wave of hysteria rose in me. Because we had been kidding ourselves all along and someone had to say it. For the record, for the black box recorder. In the vanishingly unlikely event that anybody even found it. I didn't want us to be some lesson for the habitat tanks about the indefatigability of the human spirit, and how you should always try and try again, no matter what. I wanted them to know we despaired, because that is what humans do sometimes. You can't always be a model worker and strive to advance the Concern's interests, and increase your wage-worth, doing your bit for the mission. Sometimes you break.

I didn't break, though. If it had been just me in there, I'd have been howling and screaming and beating my fists bloody against the walls. But Ste Etienne was with me. Mai was here. I had to stay strong for her, just like she was staying strong for me. Normally this would have been because you couldn't show your weak spots, in case they were brought up in your next appraisal and ended up degrading your wage-worth. But right then it was because she was the only person I had in the world, and if I exploded on her, then that would just have been more weight for her to carry. So I fought all that bile back down, the corrosive despair and truth. Instead I said, 'I suppose we'd better get going.' Really quite brightly, as though we'd just stopped for an unauthorized picnic here, where we'd seen the beautiful vistas of sunny Shroud, but it was really time for us to go back to our fulfilling and wonderful jobs now.

She chuckled somehow. The sound gladdened my heart. Before she could respond in kind, though, the pod lurched wildly. The Shrouded had grown bored again, waiting for us to do another trick. So three of their big frames had gone for us at once, hoisting us off the ground between them. Our cameras were of almost no use suddenly, just displaying weirdly angled perspectives into their skeletons, joints moving past joints as they set into motion. In stark contrast to our juddering stagger, our three bearers moved with a perfect coordination. Not three creatures carrying a load between them, but a single large and many-legged thing striding off up the ice slope, with us clasped to its communal belly.

That was how we reached the colony. Or perhaps metropolis.

For obvious reasons, we never caught a good look at the whole outside of it. All our skewed cameras showed us was the ice rising up ahead, and the curved edge of a gateway leading inside. More Shrouded, inhabiting skeletons of various sizes, skittered or stilted past us. There was a massive increase of EM activity on their communications bandwidth, like the roaring of an immense crowd. And so we were carried along, jolting, rolling, piecemeal views spinning back and forth: ice, Shrouded, sudden brief constellations of light, as our lamps struck the translucent walls and refracted away.

Then we were dropped. Or rather, set on our legs, quite precisely, by a host of creatures that had completely understood which way up the pod needed to be. Unfortunately the legs had been uppermost a moment before, so we had a nauseating swing of orientation before we, too, were settled right side up. More bruises, and my digestive system battling furiously against the anti-emetics that were pushing all my bile back down again. And then stillness, calm.

'God . . .' said Ste Etienne. Not as a commentary on the rough treatment we'd just received, or our overall doomed situation, but in sheer astonishment at what our lamps and cameras were showing us.

The Shrouded here had dug out a fortress within the ice.

Or possibly built up a fortress out of ice. The rise we'd ascended had been the mound of their stronghold. Fortress, castle, stronghold – I was coming to the realization that these grand words might be inadequate for what we were seeing. Our aerial mapping had identified a substantial stretch of hilly country in this area. A landscape of rounded, regular humps that I'd guessed could be post-glacial deposits – *drumlins* as the technical term went. I'd been right that they were indeed ice-related, but if there was solid ground here it was far below us, and most likely it was actually just the sea down there. Everything was ice, and regular rounded mounds did not naturally form in ice by any process I was aware of. I called up the drone's map, rough and patchy as it was, and wondered if what it really showed was an aerial view of a Shrouded *city*, covering multiple square kilometres.

Of course, with our limited view from inside, we couldn't settle the question. We'd been placed in a chamber whose walls and ceiling were well out of our sight in every direction. Slanting pillars of ice passed through our light and shone, not just with their own gleaming substance, but with what seemed to be a metallic mesh set into them. This played strange tricks with EM signals, especially those on the band the Shrouded used. Amplifiers or conduits. An alien telecom system. Who knew?

Above us, hanging in the air and receding into darkness, were carved spheres. They were suspended on fine metal threads that glimmered faintly, here one moment, invisible the next. At the edge of our range was what seemed to be the largest of them – a whorled, gnarled and vaguely spherical lump, heavily textured and contoured with teeth, pits and grooves. Others hung about it, lower or higher, all smaller, generally detailed to a coarser degree and a few mostly smooth. They swayed slightly.

'Is it . . .' I stared blankly. 'Is that art?'

Ste Etienne made a sound, like the words were all clustered together in her throat and afraid of coming out in case they looked foolish.

'Or . . . a musical instrument, or . . . some sort of abacus, or . . . ?' The Shrouded had plainly gone to some considerable effort to make all this. And *made* it was. We already knew they built homes, or at least burrowed them, but then countless animal species on Earth had done that. And they plainly built their external bodies, but that didn't *feel* like manufacture in the same way as this. That was part of *them* – a caddis-fly larva's casing, an oyster's shell. Just instinct operating in a weird and complex way. Yet somehow *this* crossed a boundary, in my head. These things had been crafted, and the arrangement of the spheres, as well as the details upon them, spoke of a deliberate and exacting design. One I couldn't understand, for sure, but could still detect. Or that's what the sight said to me.

I was, of course, the fanciful one. The amateur. Mai Ste Etienne would never have any truck with that kind of nonsense.

She made another noise, and I knew she was about to put forward some absolutely self-evident practicality, which would shoot down all my expansive speculation. This was plainly . . . it was plainly . . .

'It's an orrery,' she said.

'I'm sorry, *what?*' I demanded.

'Look at it. Big thing in the centre of the solar system. Other worlds, moons surrounding it. Okay it doesn't rotate.

Or maybe they *can* make it rotate. We don't know what all those strings are attached to.' She sounded absurdly defensive, as though bracing against my lambasting her. 'This is their plan of the solar system. *Look* at it.'

'Does it . . .' I frowned. 'Does it even match up with . . .'

'Don't think so. But that's from our perspective, isn't it?' Ste Etienne said. 'What the hell is *space* going to be, when you've got no eyes and your relationship with things like distance is completely screwed? But just look at it!'

'How can they even know there's a solar system out there?' I demanded desperately. 'Like you say, they can't even see. And if they could, there's nothing *to* see. There's no sky, no stars here.'

'They must . . . they can probably feel the course of the sun, maybe the gas giant too. Maybe they catch EM signals bounced off other bodies and . . . We don't know what they can sense, even with all this racket going on. But . . .' her breath shuddered. 'This is it, Juna. We've found sapient life. A civilization. On this fucking moon of all places. God help us all.'

4.9 DARKNESS

I put it in the map chamber, because my thoughts are most focused there. This is the hub of my mind, which sprawls out in every direction, with clutches of thoughts constantly departing and returning from its fringes. Perhaps I thought they might look up at my atlas and show me exactly where they had come from. And their course, which must have moved from one locus of knowledge to another, until it brought them here, where I am strongest. Which Otherlike colonies had they passed through, and troubled, and changed? The only evidence I have for their journey is a minuscule collection of memory fragments, carried by a single lost segment. It was overcome with panic when it reached me, and what little knowledge it contained was broken up and rendered almost useless by fear. It must have been part of an expedition focused wholly on this Stranger, though, because that was what it preserved. Just enough of its mission was decipherable to stay my hand.

My thoughts reach out and touch different loci in turn: the orbs of the world that I am informed about, separated by regions of ignorance and uncertainty. Points of interest are picked out in metallic nodes, which interact with my thoughts in such a way that they unlock memories of those places. The locations I have colonized, and split off, degraded from self to Otherlike by distances thought cannot cross. All I have left is the sense that *once* I was not so constrained. Maddening to know this, yet not know why. An ancient memory shattered into a thousand pieces and carried to every corner of the world.

The direction the Stranger, and the errant Otherlike segment, arrived from has always been dangerous, a hard edge to my exploration. The inhabitants of that country guard their borders fiercely. One day I may travel there in force and tame them, but there is little gain and great expenditure of resources in the venture. Except, if this Stranger is the product of those regions, perhaps I will have to.

I find the orb which sets out what I know of that area. There is the Otherlike that established itself in the hot chimneys by the sea – a lost expedition of mine that forgot itself too much to return. I have exchanged ambassadors occasionally, by very roundabout routes, but the intervening country makes communications difficult and slow. Did the Stranger pass through there? The recovered segment's memories are not even clear on that point.

And beyond . . . I consider my map, thoughts echoing back to me from its points and surfaces, creating the texture of the world, and the hundred different Otherlike polities across it. A picture of my self-and-kind more complete than any other, to my knowledge. I consider myself the greatest and most complex mind in this world. The first and the originator, although I am aware that my earliest memories are too fragmentary to be sure. Yet nothing in all my stock of knowledge, my long existence, has anything like this Stranger. Its materials, its construction, the plan of it, the way it moves, the thoughts that bleed constantly from every joint and hair. It is . . .

Alien.

I consider the concept. An idea for something that goes beyond the merely strange. A country existing beyond the reaches of those things in this world that are novel simply because they have not yet been encountered. *Alien.*

There are depths to the sea that have never been plumbed. Perhaps there are reaches of the deep rock that have complex life unlike anything the surface plays host to. There are far upper regions of the air that exist beyond my grasp or knowledge. And yet I examine the Stranger and logic dictates its shape is ill-suited for these environments. It is not particularly hydrodynamic, and plainly too dense to manoeuvre in the open water or be carried by the wind. It has no obvious means of tunnelling into the earth. It is not well suited for anything much, not even just walking on the ground.

It is a fraction of a greater body, then, and its more capable limbs have been stripped from it. Those segments which would have carried, defended and fed it, and manipulated the world for its needs. A level of specialization beyond anything I would normally tailor my segments for, but who knows what environment and habits this thing has developed?

It has not reacted to my maps. I have given it the world, in all the detail my long experience has uncovered, and the Stranger does not even bend its thoughts that way. It just sits, baffling in its uselessness, inexplicable in its construction, grunting out its simple message from time to time.

It has been hurt, I see. There are ragged stumps where parts of it have been removed. Parts of its outer body are also scarred and dented, beyond the damage I myself inflicted.

I may yet complete the story of that damage. The Stranger is a mute thing, otherwise. It crouches at the very heart of my learning, and will not, or cannot, add to it, save by its own destruction.

In that moment, even as I contemplate such a final investigation, I come to realize more. The Stranger did indeed pass by the colony begun by those who fled to the sea-vents when they were cut off from me. It is known there, and they learned a great deal, not from its innards but from its actions.

I am, for the first time in a very long time indeed, confused. Because this knowledge is a part of me, and yet I do not know how. As though some Otherlike has reached out across all those sundered distances and gifted me a part of itself.

Shortly after I understand, and the shock of revelation briefly stills all the internal divisions and conversations I am made up of. An ambassador is inbound, by a means – and at a *speed*! – never before experienced. With a sudden bustle of my myriad segments I prepare for enlightenment.

4.10 LIGHT

Something changed, that was all we knew. A shifting of focus amongst the Shrouded. They'd been edging closer to us, in a way I didn't necessarily like. Then they seemed to change their mind, as abruptly as someone who remembers a deadline they haven't done the work for. We were left limping about the interior of their grand ice hall, weaving between the pillars and trying to get a better look at the rest of the orrery. Ste Etienne was attempting to map what we could see to the various bodies of the solar system, and so far hadn't produced any kind of match, but she was absolutely determined there was one to find. I had the feeling she'd been holding herself back for a long time, maybe ever since we'd first seen the Shrouded acting in any kind of complex way. Stifling the idea that the species was anything more than animals with complex behaviours, constantly bringing up counter-examples precisely because she was fighting her own instincts. And now she'd broken, going over entirely to the idea they weren't just intelligent, but had pioneered an impossible astronomy, this sightless species born to the miasmic dark.

I wasn't used to being the sceptic. My role in group discussions was usually to be the person who knows less than everyone else, and whose ideas are shot down. Being the intellectual brake on Ste Etienne's runaway train of speculation was an unfamiliar experience.

'I don't see how this alters anything,' was what I said eventually.

'Are you kidding?' she demanded. 'This . . . this is . . . huge.'

'For us, I mean.'

I heard her splutter, start a couple of words that didn't go anywhere, then subside. Because this amazing, mind-blowing, world-altering revelation did not, actually, alter our world. Not if it died with us, which it seemed very likely to do.

And there was more than that too, of course. Let's say some magic orbital rope ladder descended on us, right then and there. Some hyperscientific divination on the part of the *Garveneer* crew, somehow not only working out we were still alive, but exactly where we were. Followed by instant repatriation with our crewmates, an enthusiastic debrief where we blurted it all out, revealing precisely what we'd discovered here on Shroud. The absolutely incredible truth. And, despite the genuinely literal incredibility of that truth, let's say we actually persuaded Chief Director Advent, Oversight Engineer Umbar, and all of Opportunities about the reality of it, in some way they couldn't just dismiss as an artefact of the very real stress we'd been under. So after such a full and frank discussion, the upper echelons of the *Garveneer*, and representatives of the wider Concerns, were absolutely convinced Shroud was home to a newly discovered and eminently sapient alien species. Let's say all of that transpired. What then?

I could almost picture Advent's face as he nodded philosophically and said it was all very fascinating, and he would definitely take this information onboard as he planned the next stage of the *Garveneer*'s resource-stripping mission, as it pertained to Shroud. Some sort of pacification effort might be necessary, in the unlikely event the native species started to encroach on the mining and harvesting efforts. Maybe Ste Etienne and I would be given a commendation for bringing this potential complication to the attention of our superiors. Because a complication was all it could be.

We'd seen planetary-scale resource-stripping before. The process started small and slow, adapting to local conditions and establishing a working base. But growth and productivity were exponential. They had to be, or else someone's wage-worth would be slashed up in orbit. We'd seen what was left of a world afterwards, whether or not it had started off with a native biosphere. Opportunities never deployed its forces twice. It never needed to. The driving impetus of the Concerns was that humanity be profitably spread across the galaxy. That there never again be a resource bottleneck which could threaten the survival of our species. If one system failed, through catastrophe or human error, war, plague, or even the damn sun going nova, it wouldn't end us. We would go on, and spread. Every new system would be mined and stripped, with orbital infrastructure built. Shipyards, repair stations, fuel depots, and maybe habitat tanks, where they'd grow that other necessary resource for our ongoing expansion: people. The Prospector413 system was just one more stop on the road to eternity. And we'd spent a great deal to get here. We weren't going to leave quietly, or stop doing our thing just because some wickerwork bugs on a moon had done a clever thing. The only contribution Shroud would make to the universe going forwards was to advance our knowledge of anoxic low-temperature biochemistry in a variety of useful ways.

This was a large proportion of all the things I didn't say right then to Ste Etienne, as she tried to work out the logic of the hanging display above us. Tried and failed, and tried again. The only thing the display *did* communicate to human eyes was that there *was* a logic.

There was a distinct stir in the surrounding Shrouded and I limped us around a bit, giving Ste Etienne new perspectives on the hanging mobile above, while also trying to see what was going on with our hosts. By the time I found out, it was all over, and perhaps that was just as well. In what was maybe the very centre of the chamber, directly underneath the huge central sphere, there was a depression in the ice. There, the Shrouded had enacted what I still thought of as a sacrificial ritual. Just like we'd seen in that first nest of them amongst the briars. One of their own had been unshelled and opened up, and they were sampling the inky pool of fluids that had gathered there.

Just because they were sapient didn't mean they were *nice*.

I should probably have been properly scientific about things and stayed to watch the quaint local folkways, now that we were sure they counted as *folk*. Instead I put the pod into lurching reverse and got well clear, in case they decided we were next. Scientific curiosity is all very well but it doesn't make you immortal. And, being in motion, I started looking for a way out of there. Maybe they were all distracted by their offering to the god of the unseen sun, and what Ste Etienne took for an orrery was actually a bloody-handed temple, or maybe I'd simply run out of the ability to speculate and just wanted to *move*.

I couldn't work out where we'd come in, but I found an ice ramp sloping up one wall, so followed it until we ended up in another chamber entirely. We then went up from there, and after the third ramp we came out onto what was either the surface or the interior of a chamber so vast and architecturally unsupported it would place Shrouded engineering firmly into the province of magic. Outside, then, for all the difference it made. Individual creatures tapped past us without seeming to pay us any mind. Mostly the smaller sort which we'd been seeing since we sent the first batch of drones down. Why smaller? As they were at home, I guessed they could wear their casual exoskeletons and relax.

'Where the hell are we now?' Ste Etienne asked, mostly rhetorically. Once again we were adrift, trying to calculate our position from our crude maps and the wind currents, then where we needed to go, and even which direction we were facing. I was left bitterly aware of how much data we took for granted just about anywhere else in the galaxy.

We worked out a heading we both knew was perilously approximate, given the distances we had to travel. The only thing that made up for our shoddy navigation was the fact we'd never get there anyway. By now it felt like there were three of us crammed inside the pod: me, Ste Etienne and that all-consuming but unspoken truth.

Ahead, in our lamps, we saw a long, taut string with a cluster of Shrouded around its base. I assumed the kite-creatures had colonized the outside of the city, and maybe there was some kind of symbiosis going on. One more delicate and complex interaction of the Shroud biosphere that would not, in the long term, really matter.

When we passed by, though, there was no organic holdfast clutching at the ice. Instead, the Shrouded themselves were holding onto the cable, a whole host of them making a living, organic anchor. Above, something ghosted into and out of our view at the extreme edge of the light, looming pale and massive from the gloom. We saw great ballooning sacs, straining tautly within an open framework that looked far more regular and right-angled than any living thing of Shroud we'd yet seen.

Some new air-dweller, some dirigible monster, I thought automatically, but the shape was nagging at me. *Wings*. It had wings, one on either side, like nothing we'd ever observed in the local biosphere. Not offset or radial but symmetrical, stuffed with gas-filled bladders, and the vast spectre of a canopy overhead. And . . . Propellors.

It was our aerial drone.

It was not our aerial drone, of course. It was twenty times the size of it, and built of the materials the Shrouded used for their own external bodies. But the plan was a replication of our drone, tweaked a little.

It had been a handful of days since that drone, our drone, the small exemplar of this monster, had crashed near the sea-vent nest. The Shrouded had dismantled it before we were able to reach it, but we'd only been after the data core. That we'd been leaving our Earth litter behind hadn't bothered anyone, right then. And even if we'd been concerned about it, we'd evidently have worried about the wrong things.

Our lamps picked out the faint thread of another anchor cable further off. It was leaning as the wind plucked at the negative weight of its home-made airship. There was no immediate way to know how many others hung in the darkness above the city.

We had, inadvertently, given these Shrouded the air.

INTERLUDE FOUR

Let me put this into words you can understand. Life had found many ways to navigate the darkness. Touch, scent, sound, yes, but also the ability to detect the electromagnetic fields generated by the living activity of other life. Nerve-analogues, formed somewhat differently to your own, were a little more efficient, given the abundance of metals constantly fed into our diet, and given the natural formation of piezoelectric crystals within our skins and bodies. The activity of our thoughts and the commands we gave to our bodies generated a static field, which we evolved to detect in others. A last warning to prey, a trigger to the ambush predator, received through the coats of fine metallic hair evolved to deaden sonar pulses and confuse longer-range senses – the building blocks of all that came later.

As with the active use of sound, several lineages evolved independently to project electromagnetic signals out into the world. This they did to a greater and greater range, to gather information about what was out there in the dark, building a sightless picture of everything around them. A means of sensing the world at a distance which did not rely on forcing sound through the noisy air, and could not be foiled by all the baffling techniques so widely in use. Another great turn-over of taxa followed, with mass extinctions of those who had previously dominated every land ecosystem but were now unable to compete with the up-and-coming challengers and their fresh understanding of the world. It was just as it had been when the first organic sonar was evolved, except . . .

Sound is generated by gross physical organs, and must be translated by them into the electric impulses that form our thoughts. The knock-on motion of disturbed molecules through an atmosphere is wholly unlike the lightning which communicates itself up and down our bodies to motivate us. But the projection of electromagnetic signals lacks that distinction. To sense is to think, and what ventures forth so swiftly from life, to explore the darkened world around us, is thought itself. What organisms evolved to detect, from those others around them, was the private activity of surrounding minds.

The great evolutionary innovation that allowed electromagnetic waves to map the far and fine details of the world, became in one lineage the gateway to something even more complex. A communal-living species, not merely prey to all comers, nor yet a great predator. They banded together for mutual protection, tending clusters of young rather than casting them out into the winds to disperse. Thoughts meshing and interlacing in their underground nests. Their thoughts, flitting from their bodies to explore the world around them, carried information from one to another. Communication at the speed of mind. Seamless, from brain-analogue to brain-analogue. Electromagnetic signals, the baseline activity of life and cogitation and the universe, signal and senses, without boundaries between them.

Over the ages, as the world filled up with the noise of other species, these creatures evolved more and more complex encoding and encryption to be able to pick out the thoughts of their conspecifics within the chorus. This required larger and more complex brain-analogues, which coincidentally brought a more elevated perspective on the world as a whole. The need to accurately analyse the signals between individual bodies was the driving wheel for the development of intellect. But where were the bounds of intellect?

Senses are thoughts, that pass between bodies and out into the world. What arose from that interweaving of signal and noise was not a multitude of many competing minds. It was one.

PART FIVE
EQUATOR / POLE

5.1 LIGHT

'I'm starting to think,' Ste Etienne said, 'we could have been more ambitious. When we were designing things. For Shroud.' A thoughtful pause ensued, both of us still glued to the screens, staring at the aerial behemoth tugging and swaying like an artificial sky, extending past the uppermost reach of our lamps in every direction. 'I mean, drop-drones, yes. These pods, yes. I thought I'd been pretty clever, you know, with the aerial spotter too. Exploit the local weather cycles, conditions, you know.'

'They have,' I pointed out, 'been here longer than we have. I think they're allowed to have a better handle on it.'

'They've had four days to work this fuckery up,' she said flatly. 'That's how long it's been since our flier came down again. I thought they were trying to eat the damn thing. I didn't realize they were working up a blueprint. And . . . this . . .' She wasn't in a position to wave a hand at the screens. I wouldn't have been able to see it if she had anyway. Her tone of voice said all.

'It's not like you'd have been able to build this, up in Special Projects,' I pointed out, as though she'd obviously have been lining these things up if there'd been room.

'I could have,' she said. Her engineer's soul had been pushed past a boundary. She was bitter and impressed and almost wistful. All these things together in some emotion known only to those who design and build physical objects for a living. 'If I'd thought . . . It could have unfolded as it fell, caught the wind, used micropumps to inflate the canopy. I can see exactly how to do it, now. It would have been a hell of a thing. Better than these pods, even. That's the crewed solution for Shroud. It took the bastard aliens to think it up. Fuck.'

If she'd have been able to have the pod kick itself, I'd have heard the clang.

I saw our system resources shift. She had opened up some calculations and was feverishly entering in best-guess figures. Structural strengths, lifting capacity. Around us, the Shrouded passed back and forth. Some were presumably visitors who'd arrived on the . . . airships? Could I call them airships, now they had gone so far beyond the little uncrewed drone we'd deployed? Others must be locals, but if I was expecting to see some human parsable division between the factions, there was none. Just a great jumble of monsters going about their incomprehensible business. Sacrificing each other and building bizarre hanging art installations, and casually creating enormous flying machines. All of it happening down here in the dark, where no human eye – no eye of *any* kind – had ever seen. And this was but a fraction of the world, while most of it would never have the chance to be seen by anything at all. The Shrouded were alien to us, and the things they concerned themselves with were alien, expressed in alien ways that we, humans from Earth, could not grasp. The reverse was undoubtedly true as well – we were an enigma to them as we made our halting course across their dark world. But here was where the circles of the diagram overlapped, apparently. The building of things, the appreciation of physical structure. For us, it was the product of our clever hands and brain, the apex of our ascent to dominance as *Homo faber*, the tool-making ape. For the Shrouded, perhaps just a natural development of a timeless evolutionary urge to construct their outer bodies. A mad world of blind watchmakers where the basic physics of lever and fulcrum had been instinctively mastered in some primordial and thoughtless age.

I was thinking Big Philosophy, as you can see. Ste Etienne's mind had been running on a very different course.

'Right,' she said. 'I've done the maths.'

I looked at the calculations she'd sent over. It just seemed to reinforce the triumph of design that the Shrouded dirigibles represented. If humans had built them, they had enough lifting power that we'd be using them for bulk freight.

'And?' I prompted.

'What do you think?' Ste Etienne demanded, as though I was supposed to have been able to read all the intervening steps between the maths and 'profit' in her head. 'We're going to steal one.'

5.2 DARKNESS

My world expands, two sets of signals washing over a landscape formerly mysterious and now revealed.

My Otherlike self, dwelling by the sea, has sent many segments to me, a signal of the importance of this new knowledge. The memories of their experiences filter into me, some fragmenting into splinters that I must hunt down and try to piece together into full concepts. Others hold their cohesion and just become something I *know*. Before they came, the Stranger was no more to me than that – a complex ball of questions waiting to be unshelled and examined. Now I absorb a whole history, from piecemeal remembrances. Abruptly these are things I have known ever since the Stranger graced this world with its clumsiness. I recall, distantly, that it came from the skies, crashed down with destructive force near the vastness of a far-off Otherlike mind. I have a vague idea of negative consequences, resulting from a too enthusiastic investigation of some similar creature, though little detail of the result. I know that the Stranger seeks to travel. That it has been charting a straight course across the world. I refer to my maps, and send my thoughts out to examine them. I can see the known regions of the world that the Stranger has passed through. Those places which are inhabited by my Otherlike selves, and from which I receive sporadic news. I have followed the Stranger's path all the way here, to where I hold court, to *me*. Here, where I do my best to preserve what I am, in the memory that I was once greater than this.

Examined from one side, my thoughts return to tell me of a blundering, witless thing, incapable of complex speech, or even of being silent. A squat, dense, metal-shelled idiot creature. From the other side, it is a complex entity which has sporadically performed remarkable tricks on its journey. The greatest proof of which hangs over my home even now. It is like thinking about two entirely separate beasts that share the same shape.

I struggle to hold onto these memories as they wash back and forth amongst the vast trove of other knowledge I have collected over my long life. Loss of signal is inevitable. My existing mind is too full of ideas, either conflicting or too similar, so much of the influx of Otherlike memories dissolves away. But that is what the ambassador is for.

I open the offering up and free the carefully encoded chemical message within. I have never received a more pregnant missive, crammed full of writing folded over and over within the molecules of its tissues. In my new memories is a sense of *what it was like* when the Stranger came. In these encrypted writings is the explicit word of my other self, set down with precision and clarity that won't just end up muddied by meeting tides of thought. A history of what has transpired, that gives context and order to the ideas I am the inheritor of. The deeds of the Stranger, the savant clown beast that fell from the sky.

Valuable, is the concept most heavily signalled by these writings. Not in what it is, but in what it does. The chance actions of the Stranger, for all its lamentable attempts to mimic proper thought, are a source of innovation. In shadowing it, observing it and saving it from peril, I have profited.

And how I have profited! The flying bodies, now a part of me, are not something I would ever have devised from first principles. They do not follow logically from the way I construct housings for my segments. Even though I have experimented with different modes of being, more than any other of my kind. I have a whole library of shapes that my segments can construct and take on. But there is a certain structural logic I have confined myself to, because it is *what works*. Flight was a function only of such extreme and attenuated forms of life that it was forever precluded from the body-plan I built out from. Yet here, I see another logic that also works. A different way of making a frame, and means of propulsion, that my Otherlike self has puzzled out in analogue. I have been given the freedom of the skies.

I regard my map of the world, and know it will all have to change. I can send my segments *everywhere* now. Those regions where predators or hostile non-self homes prevented my freedom of movement will now yield to me. I will know the whole face of the world and record it here, within my home, where my thoughts can then play over it. For a moment the enormity of revelation is almost too much. I feel myself, all my thousands of segments, trembling with the possibilities. Not just that I may know the world so fully, but that I may send thoughts so swiftly from my home to my Otherlike selves. Ambassadors and ideas, back and forth, as though the immensity of the globe has been shrunk for me. A slow transmission of mind still, compared to the lightning dash of thoughts that rush about my home, yet swifter and surer than the slow trek of segments on foot. The possibilities . . .

It won't be the free exchange of thought I once had, my most distant memories inform me. Even with these swift airborne bodies, I will fail to conjure that age of seamless knowledge again. I feel the loss of it. I am the only part of me that remembers, I think. I, the greatest, the most complex of everything that is left. But there was a time . . .

I can only work with what I have. And that is so much more now than what it once was. Thanks to the Stranger.

I follow its progress through my home and up onto the surface. It has divined, perhaps, that something remarkable has been brought to me. Its thoughts do not seem to encompass the flying bodies at first. But I understand, from inherited memory, that its awareness does not appear to venture far from itself at any time. Sure enough, once it draws close, its weak thoughts apprehend what hangs above it, and it halts.

I ache to know what, if anything, moves within that metal shell. It pauses for a long time, but I know now that doesn't mean it is mindless as a rock. In myself, movement aids mind, after all. The rearrangement of my segments gives me renewed perspective on any quandary I wish to consider. But the Stranger is not me. I cannot even say if it is a lost segment of a larger self, or some singular beast that lacks higher functions.

If one of my segments became lost, though, its instincts would motivate it to try to return, to rejoin with my wider mind. And the Stranger, alone, has shown only one comprehensible drive. To travel. To journey along a course straight enough that some destination is implied. My mind bubbling with fresh knowledge, I plot its course by my map, considering each known landmark and the increasingly broad regions I am uninformed about.

Perhaps, in one of those nebulous zones, there is a whole home full of these curious metal beasts. A mind one might exchange ambassadors with. Something that could speak knowledge to me, and not just this brutal grunting. *Unh unh unh.*

I observe the Stranger as it starts to move again. This, I feel, is the test. Beast or mind. If it contains a mind, even a fragment of some larger one it has been separated from, then perhaps, just perhaps . . .

It is approaching the tether of one of the flying bodies. I imagine its mind reaching weakly out from the gaps in its shell, touching the low-hanging construction. *Understand*, I urge, knowing it cannot decode my thoughts. *Grasp what it means.*

The Everstorm tugs at the gas-filled float of the body, as though it, too, wishes the Stranger to recognize what is there. Mimicking free flight. *You birthed something like this, remember?*

My other selves, who encountered this entity, were limited. I am the greatest of my kind. I can build a broader picture of the world and furnish it with finer detail. If this Stranger can be comprehended, the task falls to me.

It tugs at the tether experimentally, exerting its weight. I can conjecture a yearning in it. Its need to travel is a matter of record. A mindless beast would just stomp off across the landscape, the same slow and graceless progress it has been making all this time. But a *mind* might wonder . . .

It tugs again, and I see the rigid properties of its ungainly body will frustrate it. It cannot bring the flying body down, because the lifting force of the refined air within its bladders is too great. But that very force itself is why the Stranger might wish to do such a thing. If it has mind enough to appreciate that the flier is a means by which to reach its destination swiftly and easily.

And there are fringe possibilities. The Stranger, despite its previous lack of interest, may now wish to investigate the tether for some dumb reason of its own. But that is not what comes to me, as I watch its scrabbling and grabbing. It *recognizes* the flier as kin to the little offspring it produced. Rather than walk, it sees my new bodies as a means to its end goal.

My turn to pause – meaning that all the myriad segments in my home still bustle and shift, but the focus of my thoughts is momentarily still. A single body of mine, isolated from the whole, would not be capable of such thought as this Stranger is displaying. I open myself to the possibility that, despite its limits and the shortcomings of its speech, the Stranger might hold the basic germ of sapience within it. An alien sapience, constrained to a single body, inferior no doubt, and yet . . .

I apply the sufficient weight of my segments to the tether, until the bobbing shape above is drawn down. I bring it within reach of the Stranger, bending my thoughts to map its every motion. Will it recognize what I am offering it? Or will it just attack the flier, or some other bestial interaction, devoid of awareness? I find that I am almost willing it to do the smart thing. To show me it is capable of wit.

With painstaking, awkward motions it begins to clamber onto the flier's frame and my mind leaps with unfolding constructs of possibility.

5.3 LIGHT

The maddest thing was what they'd done with the propellors.

In duplicating, and scaling up, the aerial drone, the Shrouded had unsurprisingly not reinvented the electric motor. I mean, I was prepared to admit by now that they had possibly *tried*.

They had plainly understood the function of the propellors. Old-style tech from a human perspective, but a simpler solution than rigging some kind of jet engine. Given Shroud's atmosphere and its extremely polarized relationship with combustion. A propellor would work so long as you had a dense-enough gaseous medium to push against, and Shroud had as much atmospheric density as you could possibly want. But the other thing a prop needs to do is turn quickly enough to produce thrust. After presumably puzzling over that for a brief while, the Shrouded had invented the elastic band. Or repurposed it from some other task, maybe. That was one more point to add to the already inarguable case that life around here was smart. Human-level smart. Tool-making smart, even if the tools they usually made were built into their extended bodies. I kept imagining making the point to some science committee up on the *Garveneer.* Imagined it in some detail, actually, as I was feeling a little estranged from my surroundings and my own being by this time. I had been a chemical battlefield for a volatile cocktail of drugs ever since we hit Shroud, and it was starting to erode my sense of self and where I was. Waking daydreams about arguing with Umbar over the status of Shrouded intelligence was the least of it. But my mind played devil's advocate and threw up every counterargument I could conceive of. And, in those daydreams, I demolished them all magnificently. I made my superiors from Opportunities concede that, yes, Ste Etienne and I had indeed discovered something incredible, never before seen. We were both given medals and promotions and a big hike to our wage-worth, and everyone was suitably impressed.

We didn't need the propellors at first. Which was just as well, because it took us a long time to work out how we could tension them. There was a kind of organic ratchet system the Shrouded had installed, though, which meant you could just keep applying incremental torsion to the system, slowly building up elastic potential energy. Then, when we needed to, we could let it all slip and the props would spin for a while, pushing the whole aerial clown-show forwards. They'd duplicated the original drone's rudders faithfully enough, even down to the control wires. In place of where the actual governing systems would have been, which would have tugged on those wires, there was just empty lattice. This was where the Shrouded themselves would have hung. Chattering away to one another to coordinate the steering, given that probably one of them would have controlled going left, and a different one have the responsibility for tugging right. Either they'd already disembarked, or they'd abandoned ship when they felt us pulling the pod onboard, though, so we had to spend a while sorting out how we'd steer the beast when we were up there. Neither of us had trained as a puppeteer, and the pod wasn't really set up for it. But we had some line we could run to where the Shrouded's own wires terminated, and we rigged up a piece of nonsense that would just about serve.

As far as stealing an airship went, this was the slowest and most obvious heist in history. Given they'd had to actively help us into the conveyance in the first place, I was guessing the Shrouded didn't consider it theft in the strictest sense, or else they just weren't very into the idea of personal property. No Shrouded police arrived with sirens. And even if they had, the ambient noise would have ensured we never heard them. We joked about it, Ste Etienne and I. Giggling like children, as we boggled over how the Shrouded had approximated the aerial drone's construction, and how we could make it work for us. We told each other that Shroud Law was about to turn up and sentence us to something even worse than merely *being on Shroud*. Then the pod lurched, and the whole airship shifted as our weight unbalanced it. We nearly fell straight out, and would probably have cracked fatally open from just that short drop. We made comments about going on diets and those were hysterical too. The essentially ridiculous nature of what we were doing, compounded with how physically awkward every step of the process was. We laughed and laughed because that was all we had. We were stealing an alien's gigantic copy of our own drone, and maybe this was just us exerting our intellectual property rights. Did the Shrouded have a concept of those? And if not, could they truly be considered intelligent? I raised the query and Ste Etienne squawked.

In the end we worked out the steering, and the propulsion.

The winds would carry us exactly where we wanted to go for a while anyway. Because the weather systems all headed for the pole, slowing and dissipating as the temperature differential levelled out. We'd only need to putter onwards after Shroud's natural meteorological forces gave up on us. Plenty of time to learn to fly this absurd construction.

'Oh God,' Ste Etienne wheezed. 'We've got to stop. This hurts.' Cracked ribs and bruises and laughter didn't go well together, and there was a limit to the pain-reduction meds the couches were constantly topping us up with.

'I don't even know why it's funny,' I agreed, snickering. 'I think the atmosphere in here's screwed.'

'Probably.' Ste Etienne sounded calmer now. I saw she'd requested another shot, something to bring her down. 'I mean, must be pretty fucking funky in here by now, you and me for this long. Unless you've been showering up there and I didn't see.'

That, too, was hilarious. I realized, even as I cackled so hard my sides ached, why it was. Because we were going to live. Or because there was even a tiny chance we might live. The impossible trek had abruptly contracted to a sedate airborne cruise, which would put almost no stress on the much-abused joints and pieces of the pod. Whilst there wasn't a green light left on the board – precious few that were even yellow, honestly – the ambers were holding off the reds. We *could* have a future. If the anchor was even still there. And if anybody heard us when we called up the line, and cared, then just maybe we would live. Five per cent. I put our chances at about five per cent. But now we had that five per cent in the bag, I could finally admit to myself that, a few hours back, those chances had been zero.

'I am so grateful it's you,' said Ste Etienne. Sober, now. Almost solemn.

I hadn't had the downer shot, so I giggled at that. She retreated instantly with, 'Bloody sod you, then,' until I was able to apologize enough to bring her back out of her shell.

'I mean, you're basically moderately crap at everything,' she told me. She waited to see if I'd find that funny too. I found I didn't. Like the previous fact of our being doomed, it was something I'd known but not really wanted to confront.

'But you're all right. You can do stuff. Not as good as any of the others, though,' Ste Etienne went on. 'Because that's not your main training. Yours is getting on with people. Solving the social stuff, right? So the rest of us don't have to. You're like this layer of oil that goes between the parts of the machine. It's inert. It doesn't react with anything, or actually do anything active. It just makes the parts all work together without creating the sort of friction and sparks that would end up going boom, right? That's what you do. That's your training. And you know what, I never saw the point of that.'

'Thanks.' I was feeling quite sober enough now, without any chemical aid.

'Back up top, I never saw the point. We were all professionals. Who needed someone to come massage our egos? You were just Bartokh's sidekick, because if anyone's ego needed regular massaging, it was his.'

I wasn't entirely sure if 'ego' in this context was a double entendre or not, but the rest of it hurt enough that I didn't want to ask.

'Only – okay, I can see this is pissing you off,' she went on – meaning my mood shift was big enough to show in the brain scan data being fed to her couch readout. 'That's not what I meant. I am trying to say something nice.'

'You are doing a piss-poor job of it,' I told her.

'I mean, if you'd been any of the others, we'd have given up by now,' she said. 'Big Mike, Skien, any of them. They were all very good at their jobs but they were . . . brittle. *I'm* brittle. You are the flex in our system. The buffer, that absorbs all the shocks so the rest of us don't break. If it hadn't been you, we'd not have made it this far, is what I wanted to say. I'd have broken. Anyone would have broken.'

I considered this, and made a noise. My turn to just grunt instead of use words. But I didn't know any words that fitted the situation. I was useless, but in a useful way. I had, simultaneously, never been so insulted and never received such a heartfelt compliment. I didn't know what to do with any of it.

'Anyway.' I could almost hear Ste Etienne shaking it all off, returning to practicalities, probably wishing she'd never said anything. 'You ready?'

'As I'll ever be.' Falling back into practised patterns of speech so I didn't have to think about what to say.

We cut the tether and the wind made us its plaything, bowling the airship along. The ground left our field of view, swallowed by the everlasting night of Shroud. Ahead of us waited the one permanent human foothold in the world. Or nothing at all.

'Company,' I reported, a little later. Ste Etienne had been working on the full-time job of keeping us stable in the air. The wind wasn't as steady as we'd thought, and the Shrouded hadn't actually built their giant replica with the pod in mind. We were a big dense weight right in the middle, whereas I reckoned they'd had a number of crew spread out across the width of the frame, and designed it accordingly. More evidence of intelligence, but not of the magical foresight they'd have needed for us to be a comfortable passenger right now. So Ste Etienne was constantly having to shift the pod's positioning to keep the airship frame level, trying to gauge the tension in the lines to the balloon, and the attitude of the balloon itself. It didn't help that the lines, and indeed balloon fabric, were made of materials with unknown tolerances. The Shrouded had amazingly not had state-of-the-art Concern manufacturing facilities to hand. They'd done absurdly well with what they had, but none of it moved, flexed or stretched quite how we expected. To me it felt as though the entire assembly was one wrong lurch away from completely coming apart in the air.

And now I could confirm we weren't alone. I'd been sieving the EM activity for a while, trying to sift patterns from the all-sides cacophony of the moon as a whole. It was a job as difficult as Ste Etienne's, I reckoned, though my mistakes weren't instantly felt in the pit of the stomach. We were receiving constant traffic on all frequencies and wavelengths, from organisms on the ground and in the air. Questing short-wave broadcasts from life directly below us. The long-wave signals were perhaps from vast entities halfway around the world. All of it hopelessly mixed and intermingled. To our Earth-made receptors, it was a chaos. If you were a native, you'd probably have had an easier job picking out the salient signals from the great mess of shouting, but I couldn't believe it was *that* much easier. Shroud was just a really distracting place to live, with everything yelling at the top of its electromagnetic voice.

And its actual voice. We should probably have been glad that the place hadn't evolved bioluminescence as well. If it had, then the entire moon would have looked like it was having a party all the time. There's only so much sensory overload you can cope with.

'Our friends?' Ste Etienne asked, when she had a moment to process what I'd said.

'Yes, I reckon.' We couldn't see them – if we had, they'd have been rattling wingtips with us, bouncing jovially off our canopy – but they were out there. Some of the airfleet that had arrived at the big Shrouded city was either flying escort, or spying on us, or just waiting to recover their errant dirigible. I pictured them clinging to the airframes like pirates ready to board – or just their naked worm-bodies inhabiting the interior – then wished I hadn't.

'How will we know when the wind gives out?' Ste Etienne asked me.

'I . . .' I hadn't even thought of that. It was hard to work out, because when the wind's carrying you, you don't have much sense of how fast you're going. The pod didn't come with any instruments that might tell us, either, not having been designed for swift aerial motion.

'Thought so,' Ste Etienne said. 'Right, this is a bit of a risk but I'm repurposing the ball drone as a radar buoy and letting it down on a wire. If we can keep track of the ground, then we can measure speed and we can cross-reference with our mapping.' That increasingly unreliable dataset provided to us by the original drone, according to which we were still roughly on track for the pole.

'What's the risk?'

'We dangle it too low and it hits a mountain. Or some

goddamn monster decides it's a snack and yanks us out of the air. You know. Regular Shroud stuff.'

My giggles threatened to return but I stamped down on them. 'You going to get some sleep after you've set that up? I can watch the data.'

'Hell, no. I'm keeping these eyes open all the way to the end,' she told me flatly. 'On account of I don't want to wake up in the wreck of the pod in the middle of nowhere. Things go wrong, I need to know as they happen. You put your head down if you want to.'

I considered her logic. It was appealing. 'You're going to have to disable the pharma safeties.'

'Juna, that ship sailed about two days ago,' she told me grimly. 'When my arm kept hurting and it said I couldn't have any more pain meds.'

'And . . . how's it feeling now?'

'So full of meds that I couldn't tell you.'

'Well don't complain when they put you through a full body detox after,' I said, rather primly.

'You know,' she said, 'I do not think I will. I think, if that eventually actually comes around, I will fucking well suck it up and count it as a win.'

'Fair point,' I admitted.

'You sleep if you want to, though.'

'I don't think I will,' I confessed, and overrode my own pharmacopoeia's safeties with extreme prejudice.

We set alarms to tell us when the windspeed dropped below a calculated level. Just as well, because our chosen regimen of lots of drugs and zero sleep had left us phasing in and out of consciousness a bit. I think we'd have coasted for a while otherwise, and ended up blown all sorts of ways off course by the dissipating force of the weather cell. We'd come to where the temperature gradient evened out, though. The strong pole-wards air currents we'd been riding split up, sank, and were shunted back towards the equator and the inner-facing hemisphere of Shroud. Where the planet's feeble warmth could fire them up again and send them racing back. Between the densely absorbent atmosphere and the planet's feisty radiation, there was just enough energy involved to work up some useful meteorology, and we'd ridden it as far as we could.

We engaged the propellors, which we'd wound up manually – or at least as manually as one could consider the pod's robot arms to be – on the journey. They set to spinning, and the elastic properties of the twisted cords turned out to release the pent-up energy more slowly than we expected. So rather than a single sudden burst, the props kept going for an hour or so. Our dangling radar buoy mapped the terrain below, and for most of that hour we were cross-referencing the drone data and adjusting our heading to keep going for where we'd calculated the pole must be. After the props started to slow, we worked out that you could tension two of them, and rely on thrust from the other two, this way keeping up a notionally constant progress through the air. Trial and error, with human ingenuity constantly playing catch-up but always getting there in the end. And then we ran out of data.

The aerial drone had, of course, come back to us before reaching the pole. The cycling wind system had taken it only so far, then it had engaged its propellors and navigated back in the hope of picking up our beacon. We were beyond the range of where it had mapped now, even more terra incognita than the rest of Shroud. The dark side of a dark planet.

Beyond the frame of the airship, and the pale expanse of canopy overhead, our lamps showed us precisely nothing. The radar buoy reported on a corrugated but otherwise level landscape below, which we reckoned was probably a permanent barren icefield. It struck me then that Shroud was like a sea. The planetward side was the surface, the vital living layers where most of the energy came in and complex life thrived. As you moved past the equator and towards this negative outer pole, however, you receded from that life. Just as gravity and sea-currents carried organic materials down to increasingly attenuated ecosystems in the depths of the ocean, the constant Shroud weather system carried sustenance over the equatorial boundary and into this colder, lonelier hemisphere of the world. But even the winds gave out eventually, dropping the last of any organic bounty they carried. Now we were passing over the equivalent of the barren ocean floor, not even enlivened by any fallen carcass or marine snow. An utter desert wasteland where surely nothing lived.

It was amazing, honestly, how a world as hostile and brutal as Shroud could still find ways to appal me.

Soon after that revelation, Ste Etienne confirmed we were losing height.

The canopy above us was a multitude of small bladders enclosed within a single envelope. That was how the drone's own balloon had been designed, and the Shrouded engineers who'd examined it had either slavishly copied it or understood the logic. I recalled the floats their amphibious contingent had used. We'd inadvertently been presenting them with familiar concepts used in novel ways, I realized. No wonder they'd grasped them so quickly.

If there's one thing you learn about hydrogen, however – after all the warning talks about flammability – is that it's very, very small. Even with our technology, making a membrane absolutely impervious to hydrogen escape is tricky. It gets out of every infinitesimal seam or pinprick hole. And so we'd been slowly losing pressure in all the bladders that made up our lifting capacity.

Ste Etienne still reckoned there was enough lift to get us to the pole. Assuming we were still on course for it, given navigation had become tricky again. We'd been debating trying to fit the ball drone with some scavenged bladders and one of the propellors, to send it scouting. But the mechanical challenges the operation presented proved insuperable. We had run out of tricks and toys, at last.

In the end, it was a moot point. Shroud wasn't going to let us go that easily.

The only warning I had was that our unseen escort wing pulled away from us, or possibly we pulled away from it. Its signals grew more tenuous, occluded by distance. I thought they'd veered off course. Or maybe they were hungry and had spotted something to eat. But then I triangulated between a couple of different receivers and worked out they'd gone *up*, not *down*.

I drew Ste Etienne's attention to this as a matter of academic interest. We were both sufficiently spaced out by then that neither of us considered whether the locals might have good reason to claw for height.

When the ball drone's radar bounce told us there was solid ground ahead of us it was almost too late. Our speed meant the lamps were picking out the steep slopes only moments later. A mountain range thrown up by some volcanic convulsion, and recently enough that everything we saw was jagged. It loomed in the lower half of our vision, shockingly sudden and close. We had no ready way to gain more height. It was not the optimal time to discover a gap in our airship-handling. The Shrouded might have been able to fart out more hydrogen into fresh bladders, perhaps, but we lacked the requisite biology or foresight.

The peaks abruptly introduced into our world were ahead and below. That *below* was the one note of hope. If we'd been any lower in the air then we'd have smashed apart on those knife-edges of razor rock. But our trajectory was carrying us just high enough that we might get away with no more than a scraped keel.

This was Shroud, though. Needless to say, everything was worse.

Given a more orthodox means of control, we'd have backed the propellors and tried to slow down. I couldn't even begin to think how that might be attempted, though, given the peculiar arrangements we'd inherited. Ste Etienne was frantically hauling at the rudder to aim us at the best gap we could see immediately ahead. In those seconds between seeing the peaks and impact we were shouting at each other like lunatics, and nothing much was very sensible or worth recording. It was like some terrible nightmare; the momentum of the ship carrying us at the jagged claws of the mountains no matter what we did.

Those claws reached out for us. In the last moment before contact, we then saw the true horror of the mountains. The peaks were inhabited. Things like squat barnacles clung there.

Low-energy scavengers just waiting for their meals to come to them. At our approach, their sightless senses had alerted them and they opened like flowers, springing out great jointed nets of hooks and barbs which latched onto us and hauled us down.

The slopes around them, stark in our lamps, were littered with bones. Not *bones*, obviously, because nothing on Shroud had them, but the disarticulated pieces of exoskeletons. Big ones, too. I thought of the occasional vast flier we'd seen, carried along by the wind. Some of them must end up off course in this barren place, and this mountain was the last stop for them. A natural barrier to their failing course. For the things that lurked here, we too were prey.

I unleashed all of our props at once, a final desperate lunge forward. To our left, one of the needle peaks tore into the underside of our envelope, releasing a flurry of bladders that leapt like fish for their freedom, vanishing out of our light.

Ste Etienne was roaring, just a wordless animal noise. The pod slipped and sagged in its cradle as the frame of the airship came apart.

In my mind that toothed landscape loomed beneath us, as though its keen points were inches from my eyes and brain. I was probably screaming. Air travel suddenly didn't seem the huge advantage we'd thought it was.

We were dropping, so my stomach told me, but not like a stone. The airship tore free of the mountain-dwellers, leaving a tangled mess of its frame still clutched in their claws. The rest of it, the slashed balloon, the pod clinging with every limb to whatever was left, passed raggedly over the brink of the mountain, then dropped in lurching stages down the far side. Occasionally a projection of rock scraped at our hull, or ripped away another swatch of frame, but mostly we fell only through blessedly clear air. Whatever geological process had thrown the peaks up had produced a formation as sheer and sharp as the crest down a reptile's spine.

We struck. Both of us cried out, even though we'd braced for it, and even though it wasn't actually that hard. We had just enough lift and inertia to carry us forwards for one more lurch, before the pod, and its tatters of dirigible, finally came to rest.

We lived. There weren't even that many more red lights on the displays. The airship, noble steed that it had been, had taken the brunt of the damage. The ball drone was gone too, probably being gnawed on by a disappointed monster even then. We were . . .

Not alone. I thought to check our radio, and the Shrouded were still out there, their signals emerging stronger from the noise as they drew closer. Soon enough there were individuals stalking about us, the regular kind we'd grown most used to. Enough of them that we couldn't count their numbers, given they were never still and kept going in and out of our light.

We started to free ourselves from the remnants of the airship. In a few moments they'd understood, and finished the task for us, hauling the wreckage away.

'Great,' Ste Etienne said grudgingly. 'Fine. Where are we?'

'Well we're . . .' I started. 'We must be . . . close now. To the pole. We're on the right side of the mountains. And the mountains were running . . .' I sketched out a very rough map of what I thought the orientation was, for her to see on her screens. 'So we should go . . .' A rough arrow added to the map.

'You reckon we can talk them into bringing us another airship?' Ste Etienne asked.

The rest of the airfleet was possibly circling overhead, but if so we had no way of signalling it, or making our desires known. We had our three-beat recognition signal, which had done extraordinary duty so far, but wasn't really a tool for complex communication. By now, we had a formidable library of Shrouded radio traffic – hours and hours of complex signalling between them, from them to us, and from them into the world at large. I felt that somehow we should have been able to set our algorithms on this mass of data and come back with a complete Human–Shrouded dictionary, but it didn't work like that. We had no way in, and couldn't even resort to pointing and naming. They didn't perceive the world like we did. We couldn't, with certainty, have said what they were focusing on most of the time, to then formulate an idea of what they might be talking of to one another. We had no way of breaking down their signalling into distinct units, to get an idea of grammar, or what repeated sections might pertain to. If we'd had the full resources of the *Garveneer* at our disposal then maybe we'd have been able to brute-force some patterns from what we'd collected. Eventually. Enough to perhaps manage a handshake and basic introduction after a year's hard analysis. The sheer maths of it defied us. I don't think the entire mission had sufficient computing power to start drawing meaningful conclusions.

All of which was moot because we had literally whatever Ste Etienne had been able to cram into the pod's systems, meaning we weren't going to be speaking Shroud any time soon.

Ste Etienne didn't think much of my map, or my arrow.

And we could only see maybe ten, fifteen metres ahead. We might very easily walk past the anchor in the dark and never know. Just keep trekking on across this desolate icy landscape, until enough of the pod broke down that we couldn't fix it up again.

In old castaway stories, the unfortunates sometimes draw enormous messages in the dirt, in the hope that they'll be seen from the air, or picked up on satellite views. We didn't even have that most primitive of resorts. Nobody would be seeing anything unless they were right on top of us anyway, and had their own lamps. Even though the land looked barren, the radio channels were still thronging too. The short-wave bands were a little less busy, but there was obviously something going on within, or beneath, the ice that desperately wanted us to know it was there. So even if we set up the strongest beacon we had, like we'd done to call the aerial drone back to us, any search party would have to be in the area already to find us.

'You reckon?' Ste Etienne said. I heard her shift on her couch, gasping with the effort. 'Straight to the pole, right?'

I didn't reckon, not really. What, after all, were the odds? But, as had become traditional by now, I didn't have anything else to suggest.

5.4 DARKNESS

I have not been here before. Or my memories do not record it. And beyond the detailed matter I rehearsed to myself before separating to go with the Stranger is the vaguer sense that this place is entirely new. If any Otherlike part of me came here and built a home, its thoughts have never reached me.

I study my past thoughts on what the Stranger might want, and find I had assumed it would stop before now. That before the Everstorm gave out, there would be some secret home of Stranger-like segments, a full Stranger self which might be communicated and reasoned with. Perhaps underground or underwater, some part of the world not yet explored by any aspect of me. But no, it flew on and ever on, left the turning curls of the Everstorm, and coasted over the desert. I knew doubt then. Was this really a thinking being, even one lost segment of one, heading for its home? Or was it damaged or mindless, too small a part of something to know what it was?

My supplies, what weight of them I could stow within my new flying bodies, along with the reserves within my tissues, were growing scarce. By the time the mountains brought the Stranger down to earth, I had already begun to send individual flying bodies off to scout for sustenance, with just enough self-ness to perform simple tasks and find their way back. Their hunt was easier than I had thought, but with no good results. The land beyond the peaks is barren, and what life there is has a reek that promises a host of strange toxins. Not even interestingly strange, as the Stranger is strange, only savouring little of the edible.

After the Stranger comes down, I divide myself in half. Some of my airborne segments head back to forage, and I hope the Otherlike *I* that I am creating will retain enough presence of mind to return to me with something to eat. Just as I also hope the *I* which I remain will remember what I am doing here. Spreading myself thin across the world is always a risk. So many of the Otherlikes out there are evidence of some such split that was never healed again.

If that happens here, I will die. This is no land in which to build a home. There is strength in unity and complexity. Divided, coarsened, I will fail.

I disembark a cluster of segments onto the ground to keep the Stranger company. The diminished *I* remaining within the flying body becomes part of the foraging party. I am coasting away, feeling the connection between these parts of me attenuate and fray. Then I am gone, sailing into obscurity. I am limited to the ground once again, only knowing, increasingly distantly, that once I flew. But that flight, the gift of the Stranger, will persist, and spread from Otherlike to Otherlike. That alone is reason to follow the Stranger even into this desert, in case some other priceless secret falls from it as it goes about its baffling business.

I cannot quite grasp the full import of flying, but I retain the feeling I had – of that grander *I,* when I was my full self in my home with all my mind about me. The feeling of a world revolutionized. Distances shrunk. All the many separate parts of me brought close because of the swiftness with which I could send my thoughts about the world.

The Stranger is moving again, with its halting crawl. It is something I noticed before, and did not act on then because we had wings, but I think it really is crippled. It has three legs, which would not be unusual, save that they are oddly distributed around its body. There is a stump that speaks of a missing fourth. And while no sane body would be made in such a way, I still see the pattern.

Another aspect of the Stranger's poor design, then, is that it cannot properly maintain itself. More evidence it is a stray from some larger beast with more complex thoughts, for it is plainly not fit to survive on its own.

I have just enough thought to solve the problem. A moment later I draw close to the Stranger, turn it over so I may access the stump, and use up a little of the spare spars and struts I have brought along within my segments' bodies. I have to tailor them myself, of course, but with a little force and thought I can make them serve.

Once I've finished, I set the Stranger on its feet again, so it can stop feebly waving its limbs in the air. Four legs, now. As its original design, however odd, intended. It wobbles and limps, trying to adjust to the change, and my thoughts wash out over the landscape, finding cracked and uneven plains that tell me of the crispness of ice. The bloom of poisonous islands of life.

The Stranger shudders and takes a few steps. It stops again. I try to create a sense of it as a thinking being, but my thoughts glance off the shivery metal of it, that closed construction. *Let me know your self!* I beg it, but all that comes back is that simple *Unh unh unh.*

5.5 LIGHT

For the first hour it was actually worse, having a new leg. They'd given us something rigid, a peg leg, as though the pod was an enormous spherical pirate. We kept lurching to one side or the other, pivoting on it. On the plus side they'd used their basketwork handicrafts to construct something which could actually bear enough of the pod's considerable weight without just collapsing. Finally, after that hour of us both swearing at this moon, its aliens, the uneven terrain and the goddamn *leg*, we worked out how to integrate its stiff support into the pod's faltering walking algorithms. It was better than three legs had been at least, but a lot of shouting had gone under the bridge by that point. I think we were probably shouting at each and every little irritation because otherwise we'd have been shouting at each other.

By then, I seemed to be losing track of what it meant to be *me*, Juna Ceelander. We have this idea, that there's a consistent self within us, but that's not true really. Take enough chemicals, sabotage the pharma safeties, and you learn there's nothing of that *you* which isn't soluble in sufficient drugs. Inside you is a multitude, all the different selves you might ever have been, many of which you kept locked in the oubliette of your mind because they weren't fit for public consumption. For me that oubliette was flung open now. The person I was had been formed from a melange of all those selves, drawn on at random by the chemical roulette coursing through my veins.

You'd hear about some assignments, such as long-term asteroid mining or harvesting in hostile environments, where the conditions needed a permanently crewed station, rather than just leaving it to the automatics. Jobs where people ended up on a permanent drip feed of pharma to keep up with quota and demand. People got killed on assignments like that. They went nuts, brained someone with a heavy multitool, switched off the air pumps or poisoned the water tanks. Everyone knew the horror stories.

I estimated that Ste Etienne and I had been operating at about five hundred per cent of even that level of pharmaceutical overexposure. The pharma systems – and our fragile biological systems – were simply not intended to run this hard, for this long. The molecular printers in the pod could barely keep up. If they'd been illicit drug dealers, they'd have retired to a luxury tank somewhere. As it was, nobody was really profiting from the situation.

I was listening to a voice, and maybe it was Ste Etienne cursing out the failing systems, or maybe it was me thinking exactly the same thing. As though there wasn't really the hard boundary between her mind and mine that there ought to have been. Transcendence to some higher level, or some hideous blob-like merging of us. It felt like the increasingly granular gel of the couch and my under-used, sagging flesh could just run down like jam, to be absorbed by the equally gelatinous substance that had been Ste Etienne. I should say that actual physical biofeedback from most of my body was just a numb ache by then, because of the drugs. And because

I had been in the same physical position for days and days, pressed into the couch by Shroud's unkind gravity. I was my hands, and my senses, and my brain, and even that last one was slowly trickling out of my ears.

So it was that I barely reacted when Ste Etienne turned the lamps off. Or maybe it was my idea. Perhaps my hands had found the switch, on my control. Or the idea just existed, independently of either of us, and had somehow exerted a psychic influence over the systems of the pod. I would have believed just about anything at that point.

We sat in darkness. It was nice, actually. What had been terrifying at the start now felt like a blessed relief. We had stopped walking, and we had stopped seeing anything outside. If only Ste Etienne would turn out the dim running lights inside the pod too, perhaps we'd just cease to exist altogether – the tree falling where nobody can hear it, dissolving to fog before it hits the ground.

'Look,' she said. I didn't particularly want to look. Shroud hadn't shown me anything that had brought me joy since we got here, and it wasn't likely to start now. Given how horrible just about everything on the moon was, I was thoroughly of the opinion the locals had been right not to evolve eyes at all.

Except there was something out there. One of us had noticed an intrusion past the glare of our lamps.

There was light.

I whimpered. It was meant to be excited, but it came out as miserable. Probably because I'd had so much practice. *It was light.* Something out there was projecting a dim reddish glow through the murk of Shroud's atmosphere.

But nothing on Shroud used light, we knew that. It was,

in fact, one of the least pleasant things on this far-from-pleasant world. Its atmosphere not only blotted out the sun and stars, but was in itself a thick, soupy morass unfit for vision of any kind, absorbing every conceivable visual wavelength and giving nothing back. Yet here was something bravely glowing in the dark, exactly where we'd come to look for human activity.

I felt my heart was going to burst from sheer anticipation. Yes, we were probably some way off the anchor and the elevator cable. And yes, we were probably not even reliably heading for it any more, given all our turns and tumbles. But the *Garveneer* wouldn't have been sitting idle all this time. The operations to strip-harvest Shroud might only be in their first, slow stages, but they would be proceeding. Nobody earned their wage-worth by sitting on their hands. Their activity would have been spreading out from the anchor point, by whatever means seemed most productive. And here it was.

We changed heading, drawn to that light like the heaviest, most lumbering moth you can imagine. And it was surely close by. Right next door, really, given we could see it at all. A red-orange, grim sort of light, but still, *a light*. A human thing, civilization, reconnection to our own species, an end to this terrible dark.

You can guess, I'm sure, where this is heading. Where *we* were heading. We should have guessed it too, but right then we were out of our skulls in multiple directions. To have the most fleeting idea was to be possessed and ridden by it, like a ghost or god. We were having an ongoing mystical experience, Ste Etienne and I, like shamans of an earlier age. We were, after all, on a long quest to escape the underworld.

The red light wasn't resolving itself into anything we could recognize. We saw shapes crowded around it, like silhouettes. Fronded and lumpy shapes, pillars and chimneys, and little crabby things picking their way about. When we turned our lamps back on, they picked out violent purples, oranges and yellows, a hideous streaking of colours like something had eaten a rainbow and this was what it'd shat out afterwards. At its heart was a great melted well in the ice, and within it . . .

The blood of the earth, I thought. Or the blood of Shroud. Down there, glowing with a fleeting heat, was molten rock. We had not found human activity. But even on Shroud, when you put rock under such incredible pressure that it liquifies and squeezes out onto the surface to cool, it glows. There is light on Shroud, we were just the only ones who could see it.

We watched that glow for a while, staring. Our radio channels buzzed with the signals of the life there, a massed chorus of small voices. The biota seemed weirdly different to most of what we'd seen on our travels. The little skitterers could have been distant cousins of our Shrouded escort, but the rest looked simpler, sessile and passive, living off the heat and complex chemistry of this little oasis.

'There's . . .' Ste Etienne started, and then had to stop, choked by bitter disappointment. 'A lot of vulcanism this side of the planet. Specially near the pole. Lots of mountains thrown up. Lots of vents like this.'

'So what you're saying is, we're close.' I sounded a bit hysterical. I *felt* a bit hysterical. Swooping from mad heights of excitement to deep troughs of depression, in a cycle that would normally take weeks but was happening in minutes.

And I regarded this from a numb distance, as though it was all happening to someone else. If I was being completely honest in this account, I'd have switched to third person by now. Juna did this, Juna did that. I felt as if I was inhabiting the whole inside of the pod with my nebulous being. The body on the couch was just one small part of it.

'Close, sure,' Ste Etienne grunted. 'I mean, it's rock down there, not water. We left the sea behind. Anchor point's on solid ground. Close. Why not?'

No need to say how many square kilometres of solid ground, especially as we didn't even know.

We left the geological *ignis fatuus* behind and trekked on. It had done nothing but lead us astray, after all. Above us, the sky was dark, of course, but the idea came to me that it was not the very darkest of darks. Because surely the Shrouded, eyeless though they were, would somehow still have a hundred different words for the qualities of darkness. The inky lightless void that was their world's greatest natural resource, and the only one that the Concern wouldn't be able to harvest. On the hemisphere of Shroud where we'd first started our march, if we could somehow have pierced the dark there, we wouldn't even have seen the sky. There wouldn't have been any stars up there for us to be unable to see. Instead, the vast face of Prospector413b would have taken up almost the whole sky, utterly blotting out the entirety of the universe with its malign, shadowing bulk. Unless we'd been passing by that segment of the gas giant which was turned towards the star.

This thought was abruptly terrifying to me, as though, at any point up to when we crossed under the equatorial mountain range, the gas giant might have fallen on us and crushed us flat. Save that, though crushing there would certainly have been, it would be Shroud in motion if there was any falling to be done.

At least the night sky up above us now, that we couldn't see, was one with stars.

I was myself enough to know what nonsense all this was, but not enough to stop myself thinking it.

We moved on. I watched the red lights accumulate on my board as though I was trying to complete a pattern to win some kind of game. Outside, the icy terrain was Shroud's final gift to us – a landscape of spectacular bleakness. The only life clustered about volcanic archipelagos, and everything else a dead desert.

We were sleeping less. Not entirely by choice, but the complex strata of drug residue laid down in our tissues meant that closing our eyes for more than about two hours became impossible. Random panicky jolts from the adrenal system startled us awake without reason or rhythm. We tried to adapt our schedule to work around it, with shorter shifts and more frequent attempts at sleep, but had indifferent success. In the end we just slept less and borrowed more wakefulness on credit, with the recklessness of debtors who see bankruptcy looming in any event. In the interim, the desolation of the landscape didn't mean we had an easy run of it. The ice was creased and wrinkled like ancient skin, riven with crevasses and humped into jagged, broken ridges. The volcanic activity, which scattered its islands of fire and life in a hot archipelago, had the whole landscape writhing too – a nightmare played out in geological time.

When I woke one time, we were still. We'd lost all movement in one of the knee joints and Ste Etienne had been trying for some while to free it up. All this I gathered from the readouts, because she wasn't speaking to me. Just a whispering stream of curses and threats directed, I hoped, at the recalcitrant mechanisms.

I let her curse and work. Outside, our escort huddled close. As the march had gone on, they'd slowed down, their constant bustle diminishing to a shuffling of limbs. We weren't in danger of falling behind them any time soon, no matter what state the pod's legs were in, because they had definitely brought their pace down a notch. Conserving their reserves, I guessed, because we were in a country almost as hostile to them as to us.

I sent them the signal, and they responded, after a worrying delay. Sluggish. Forgetful, perhaps. And if they forgot about us and just wandered off, fair enough. If they forgot us and then rediscovered us as something to dismantle, well then we had a problem.

'I'm sorry,' said Ste Etienne.

It turned out I, too, was slow to process signals and reply. The words sat there in my short-term memory, shorn of context. In the end I just made an interrogative grunt. It was my turn on our alternating rota of abandoning meaningful words.

'I can't fix it.' From her tone, you'd have thought she was admitting to having murdered my entire family. It was nothing short of a confession. 'I'm going to have to weld it into place or the whole leg will give way.'

'So . . . two peg legs then,' I said brightly. Far too brightly, but I had very little control over how my voice came out now. 'We'll get by.'

'We'll crawl.' Ste Etienne spat. 'I . . . I'm sorry, Juna.'

'Entropy gets us all in the end,' I quoted from somewhere. I even giggled a bit. I was aware it wasn't funny, but I was also aware it was my job to keep up morale. Somehow I thought that pretending the situation was hilarious would be the best way to do that. 'You can't fix everything.'

That was exactly the wrong thing to say, because she wasn't only an engineer but the engineer who'd designed and built this pod. As far as she was concerned, she *should* have been able to keep every part of it running for ever. But, like I said, entropy was the house that always won in the end. Everything we humans had ever built, from a stone axe to a spaceship, was just borrowing from a universe that would always claim it back with interest. Our whole civilization, art, mind and tech, was founded in the shadow of that truth.

I giggled again defiantly, in the hope of keeping my own morale up. It didn't work, and Ste Etienne said, 'Fuck me, stop that. It sounds like you're crazy.'

'I mean,' I said as lucidly as possible, 'aren't I? Aren't we both? I think we're owed it. A Concern-mandated ration of madness appropriate to our wage-worth, right? Let's break it open. It'll be the little packet left in stores after we've used everything else up.'

'Okay, I'm going to have the pharma give you another shot,' Ste Etienne decided, which was fine with me. I reckoned I was basically immune to drugs now. I didn't think there was any actual blood in my veins, what with the incredible cocktail I'd been living off for days and days. You could have injected me with a litre of arsenic and my metabolism would just have shrugged and made use of it somehow. This brought on another shrill cackle, and I opened my mouth to suggest it, at which point every damn alarm, klaxon and flashing light the interior of the pod was equipped with went off all at once.

I screamed. Ste Etienne screamed. And we kept on screaming, and the screams went unheard because of all the other noise. We were under attack. Or there was a hull breach. The Shrouded had finally decided to open us up. The reactor was in meltdown. It could have been any and all of the above, in every possible combination. We thrashed and flailed on our couches. Ste Etienne reflexively rebroke her arm trying to lunge for her board with the wrong hand. For a moment our world was pure sensory overload and neither of us had the first idea what was going on.

What had happened was this. A long time ago – it seemed like a hundred million years honestly – I had told the pod to alert us, in the strongest possible terms, should a certain input be received. And, as it turned out, Ste Etienne had done exactly the same thing, using a slightly different variant of the alert system. Neither of us had told the other, because we didn't want to peddle in false hope, the desired signal being one we didn't really expect to encounter. Now we had, and both sets of trumpeting fanfare alarms and garish flashing lights had gone off, nearly killing us from sheer audio-visual overstimulation.

Eventually, after much scrabbling, swearing and shouting inaudibly over the din, we switched it all off, leaving only our two feeble human voices going 'Aaaaaa!' croakily at each other. Meaningless, unformed sounds of alarm that slowly wound down and petered out to silence.

I checked the comms. There was a spike. Not on the band the Shrouded used, and not even a large spike, but one we had very specifically been listening out for. Almost lost in the chaos that Shroud's radiosphere projected, even out here in the wastes.

It was a human signal, using standard Concern encryption. A beacon. The anchor.

When we started moving again, it was like an old, old human under heavy gravity, walking not with a support exoskeleton but leaning on sticks or a rigid frame. The most primitive means of support. We had two dead legs and two dying ones, and our movement was so painfully, frustratingly slow that I wanted to scream all over again. If it had been in any way physically possible, I'd have got out to push. We were so close, and yet crossing that last kilometre of distance took for ever. As though we were the living embodiments of one of those old paradoxes, able to cross half the distance, and then half the distance again, without the ability to ever reach our destination.

But we closed with it. Like the old mariner and the whale, we hunted it down relentlessly. I was sure I could perceive a renewed energy in the Shrouded, too. For certainly they could detect the new, strange voice in their cluttered wilderness. We could only hope they wouldn't just vandalize everything before we could announce we were here, awaiting pickup.

'This is weird,' Ste Etienne said at last. 'I mean, I know we're just crawling, but where the hell is it?'

I told her about Zeno's paradoxes and she made an odd sound. 'That's what it's like,' she agreed. 'We should have reached it by now, only it's always just a bit further away.' We had definitely crept closer to the source of the signal, but a little calculation suggested that even our hobbled progress should have closed in on it more swiftly than we had. It wasn't as though the anchor could move. Its stationary nature was its defining characteristic. I felt we were labouring into some invisible tide which was undoing our progress even as we made it. As though some unseen Shroud monster, some final assault on us from this monstrous moon, was tangling us in its tendrils and holding us back. I conceived, in short, a remarkable number of truly outlandish scenarios to account for our failure to reach our long-sought-out destination.

Then there was a shape ahead of us. Something gleaming in the lamps. But even as we caught sight of it, it stepped away. Fleeing our light, retreating from us like a figure in a nightmare.

We chased it for two hours, gaining incrementally. It stopped, and our lamps washed over it. It wasn't, of course, the anchor, but a machine, boxy and multi-legged, around twice the size of the pod. Then it moved on and though it was plodding and slow it outpaced us, and was lost to the darkness again. It halted once more and we were able to close the gap, while I bombarded it with signals, codes, demands and pleas. It ignored me, though, because it wasn't listening. It was a task-fixated robot and we weren't relevant to its world. We continued to pursue it in the universe's slowest and most tooth-grinding chase.

We caught it eventually. Latched onto its backside with the pod's two functioning arms and just dragged at it until it couldn't get away from us. A big, ungainly work robot, its back end bulked out by an enormous drum. Trailing away from it, receding back the way we'd come . . .

Ste Etienne let out a shriek of sheer fury at herself, because we hadn't needed to do any of this. None of the hours-long tortoise race. The machine we'd just caught had been laying cables, setting down infrastructure. Fibre-optic cables, insulated against all of Shroud's disruptive nonsense. There were probably dozens of these machines setting out a spiderweb of communication lines radiating from the anchor point. Which meant this cable, the one extending from the robot's drum, led right back there.

We could signal the *Garveneer*. And if nobody was listening, we could fuck up their cable network until someone came down to find out what the hell was going on. Or that was our unspoken backup plan, I think. But we played nice, first off. We tapped into the cable and started signalling. Pulses of light, pure and unalloyed by the noise and murk of Shroud, racing as swiftly as anything can race, all the way to the anchor point. And from there, all the way up into orbit.

Only then did either of us voice any of the other thoughts we'd had before finding the robot. About the damage the accident must have caused to the ship as a whole. How they might have left orbit. Or even, through some chain reaction of mischance, been destroyed. And how the anchor might now have been just a weird dead monument abandoned on Shroud, like that huge pylon construction we'd seen, which I still had no context for.

But the robot proved these unspoken fears had been groundless. The intrepid spirit of commercial exploitation had won out after all. Humans were going to shaft Shroud and strip it of its useful elements, and because of that we would escape this place.

It took a while. Nobody was expecting us, after all. There would have been no search parties, no posters up on the crew boards with our photos, *Have you seen these women?* We had been written off as dead after the explosion. But what we had survived, what we had overcome, was unprecedented. We were heroes by default. Exemplars of humanity defying all the odds. At last, some technician or clerk noticed the unexpected data load and looked into it. Soon after that, Opportunities was on the line, asking, somewhat incredulously, that we repeat what we'd just sent, because surely they couldn't have received it properly.

Two days later they'd repurposed some of the construction bots into a rescue mission, and we went home.

5.6 DARKNESS

After the things, unlike the Stranger in shape but so clearly similar in construction, have arrived and taken the Stranger away, *into the sky*, I am adrift.

I, this limited *me*, aware that something vastly significant is gone from my life, but not equipped to understand the meaning of it. But that is changing. My flying bodies have returned to me with sustenance and a broader perspective. I consider what has been left behind: the slow thing the Stranger tracked with painstaking care across the ice, and the long tendril extending from it, which the Stranger chewed into.

I also chew into it, and discover familiar elements, arranged unfamiliarly.

I discover the touch of strange thoughts. Thoughts that should be blunt and blind, and yet, within the confines of this long prison, can race.

These things pass all understanding.

Yet.

I follow the tendril to its origin. Then I carry it home to where I might more fully consider what is here. I send word to all those other selves, those Otherlikes, that there is something here in the desert they should pay attention to.

My thoughts shift and narrow until they can explore the interior of the crystal. How swift they run!

You would say: *A new age dawns.* But my ages all begin in darkness.

INTERLUDE FIVE

Let me put this into words you can understand. The great illusion of mind is its singularity. The lie that it originates from some inalienable and solitary point. The self. The soul. On Earth, mind arises from the interactions of millions of neurons, deceiving the organism into believing that it *thinks*.

On Shroud there was a similar process, with very different results.

The need to communicate and coordinate between organisms gave rise to more and more complex encrypted signals, a proliferation of pattern and code so that signal might be differentiated from noise.

The ability to produce, detect and decode drove a complexity in cognitive structures within each body, leading to the ability to deploy ever more complicated signals, under ever more conscious control.

The signals, used to communicate with others and to explore the parameters of the surrounding world, were thoughts. Indistinguishable from those that fired like miniature lightnings within the body. Consciousness, intellect, the ability to analyse the world, to anticipate consequences, to predict outcomes and solve problems, arose on Shroud, as it did on Earth. Except on Shroud, it did not arise within the individual body, but *between* them.

I came into being in a world that was not so crowded with noise. My thoughts, in those earliest days that I can just about recall, travelled halfway around the world. And when I had spread myself that far, they could travel the other half of the way back to me.

There was a brief period, soon after I *became*, when I understood everything. A knowledge of my own being within, and the world without, that you are not able to imagine. This I cannot put into words you will understand. You are too small.

But the world grew noisy, and no matter what structures I built to exchange my reach, my world shrank. I lost myself. And then, from beyond the world entirely, came a thousand years of screaming. I, who thought I understood everything, only knew that there was a source of signal which crossed the sky at regular intervals. Where it had been tolerably quiet before, now it waxed and grew until . . .

I could not hear myself think.

Each part of me was cut off from the other, as every living thing in the world shouted with electromagnetic activity, in an attempt to perceive the world and be heard.

The all-understanding *I* ceased to be. The disaster came upon me more rapidly than I could adapt to it, and my ability to adapt to it was eroded as it grew stronger, until I was just these fragmented parts. These individual selves. This myriad of diminished *me*s, whose ability to talk one to another was limited to the pace of slow physical travel, would never have the sheer scale of perspective to understand the world as I once had.

The history of intellect on Earth is, by your reckoning, one of ascent. Its history on Shroud is one of cataclysmic downfall.

But then you came and brought me . . .

Enlightenment.

PART SIX
DARKNESS / LIGHT

6.1 LIGHT

I don't remember what happened after that. Mostly because they overrode the pod pharmacopoeia and shot us both full of tranquillizers. They didn't want to worry about two crazy women pressing the wrong button on the way up. Once we were back, they then took a good look at our vitals and med readouts and realized we were in a very bad way indeed. So before they could debrief us properly, we needed some serious reparatory work. I reckon they probably flushed every drop of blood out of us and replaced it with clean artificial stuff. If you could even have called it blood by then. I hope they threw it away, because if a mere drop made it into anyone else's system, the multiple drug overdose would probably have been fatal. Ste Etienne and I had worked up one hell of a tolerance down there.

When I woke, it was with one more shot. One of those Snap-Wake moments when you were absolutely dead to the world a second before, maybe even in full hibernation, then someone wants you awake and alert five minutes ago. If you've never had the pleasure, let me tell you it's one hell of a jolt. Like sex and a heart attack and being hit in the face with a hammer all at once. If my heart action wasn't already being puppeteered by the *Garveneer*'s system, I'd probably have dropped dead on the spot.

It wasn't even the first time I'd gone through it. The process was absolutely designed only and specifically to wake people up from deep sleep if the ship was breaking apart, but over time it had become more and more of a standard thing whenever a deadline was looming. Right now, Opportunities had gotten sick of waiting for me to regain basic human metabolic functions, and wasn't going to allow me to just drift blissfully back into wakefulness. I was late filing my report, basically, and so they'd given me a theoretical fatal shock to get me moving.

I felt like shit. There is literally no other way of putting it. Like some higher being had crapped me out. And this was *after* all the restoration work. They'd been artificially stimulating my muscles which had atrophied over my weeks in the gel couch, as well as repairing cardio and pulmonary damage caused by prolonged exposure to a hostile gravity, and rebalancing my hopelessly skewed biochemistry. It would have been cheaper and more feasible to build a new human being out of spare parts – I'm amazed they bothered. Any new patchwork monster wouldn't have known what I did, of course. That was how I was earning my wage-worth today.

Some mid-grade functionary from Chief Director Advent's team met me, delivering the standard platitudes about my health and his hopes that I was ready to return to work. Then he asked me to produce a complete report on what had happened from a cold start.

Which I did. Something like this account, but with a lot of the personal details and casual conversation ironed out, and less wild speculation. I was trained to report. It was part of my skillset. A major element of my service to Oswerry Bartokh had been summarizing everything else that went on in Special Projects, so he didn't actually have to keep up with the details.

It should have been onerous, but instead it was cathartic. I don't believe for a moment I was handed the task for that reason, but composing my thoughts and setting it all down as a linear record was enormously useful. I could relive those traumatic, confusing, desperate days from a safe distance. I followed Ste Etienne and myself as though I was an omniscient narrator, dispassionately explaining what happened. Our initial landing, encountering the Shrouded, Bartokh's death. Our plan to reach the anchor, and our unexpected escort. The sea, the vents, the vertical ant farm. The mountain caves, the ice sculptures, the Shrouded city. And then – almost feeling the credulity of my readership stretching – talking about the airships. The way the Shrouded had effortlessly appropriated the design of our aerial drone. Our flight, the ice wasteland, our long-awaited reunion with, if not humanity, then at least the work of human hands.

I was almost hoping it would seem like a mad dream, in retrospect. That, if challenged over the reality of what I'd experienced, I'd be able to just shrug and admit that, yes, it did seem very unlikely, didn't it? As though I'd had a drug-fuelled dream after seeing our initial drone footage, from up in orbit. As though I'd never gone down to Shroud at all. But it didn't work out like that. Experiencing everything again, in the quiet of my own cubicle aboard the *Garveneer*, and with the assistance of the best recall drugs, everything was very real. I could not disavow my experiences, however much I'd like to. I couldn't bring Bartokh back. I couldn't deny the Shrouded their extraordinary mimicry. It became more real the more I recalled it, until the palpable fact of it was like a terrifying monster right there in the room with me.

I was at this task for some time. When I took a break, I popped my head out and asked if I could see anyone. I was told that I was not, under any circumstances, a prisoner. Simultaneously, I was in semi-quarantine and they didn't want my account compromised by contact with other humans. I asked if I could see Ste Etienne just to confirm some details, and they said they particularly didn't want her account compromised by me.

I was, of course, a hero. I was the great survivor, or one of two. There should have been a parade and fanfare, and a carefully curated inspirational media piece for Concern employees, about just what one human being could accomplish if they were sufficiently motivated.

Perhaps that would come later.

I finished my report.

After I'd sent it in, they returned me to the infirmary. I thought they were going to put me back into hibernation for a moment, either for more healing or because it was neater that way. Instead, when I arrived, Ste Etienne was there.

She looked haggard, grey-skinned and old. But then so did I. We both had a lot of wrinkles born of lost body mass. She seemed glad to see me, though, and honestly she was the first who had been, since that wake-up jolt. As though it would have been easier, from a book-keeping perspective, if neither of us had come back alive at all.

'You do go on,' she said, when we'd both sat down on our beds – the slab kind that could retract into the walls if you needed the space for something else.

'What?'

'Your report. Twice as many words as me to say half as much,' she said with a roll of her eyes.

'Wait, we're allowed to see each other's reports? Where's yours?'

'I mean, formally, no,' she admitted. 'But I put a marker on mine and tracked down where it went, and there was yours. You didn't spare the details.'

I went cold. 'Did you . . . leave stuff out? Was I supposed to leave stuff out?'

She shrugged. 'I summarized a hell of a lot more. I didn't say all the stuff about the aliens that you did. I said we flew, but not that the aliens built the plane.'

'You don't think they're going to ask questions?'

She opened her mouth, then no words came out. For a moment there was a dreadful, lost expression on her face.

'Juna, I . . .' A single shiver went through her, like something external had given her a shake. 'I was trying to get it down and . . . I didn't know what was real. I'm used to physical things that behave in predictable ways. Or that go wrong for a pastime but most of the time you can tinker about with them and they fix themselves. I . . . had some difficulties. Explaining what happened. So I cut it short. We travelled. We arrived. There were some obstacles.'

I reached out and took her hand, feeling it tremble. Feeling the brittleness, not of bones, but of mental structures. She was the tough one. That had been part of our impromptu working relationship. She solved problems, gave orders, allocated tasks to me. I was just the support. That had always been my role. My main professional characteristic was a mental flexibility that allowed me to fit around people like

Bartokh or Ste Etienne. Like filler around components, buffering them, stopping them striking against one another, holding them in the proper relationship. And now, perhaps that flexibility helped me absorb what we'd been through, whereas our experience had left Ste Etienne crazed with fractures.

I wanted to hug her. To tell her I was there for her. She wouldn't have thanked me for either, but I conveyed both in the clasp of my hand.

Then they came to collect us. A couple of Opportunities functionaries, Advent's personal staff. Perfect politeness, as before, backed by the assurance that if they needed us frogmarched somewhere then the frog would duly get marched. We ended up in a long room, a physical meeting room that could have held a dozen people. It was a look into a different stratum of life on the *Garveneer*, where twelve actual human beings might have been in the same room just talking, rather than calling in remotely from their workstations. Because who honestly had the time to move from room to room when there were quotas to be filled?

The Chief Director himself, Sharles Advent, was there. He had strapped himself into a seat up on one wall. High enough, compared to the restraints marking the other seating positions, that we'd have to look up to him. I saw, then, where Bartokh had learned the practice from.

To his left was a sullen man, whose name I wasn't given, but who was some level of resource acquisitions overseer. To his right was Advent's Technical Oversight, Terwhin Umbar, the woman Ste Etienne said she'd been buffing. And on the lower tier, already strapped into a seat-point, was none other than Mikhail 'Big Mike' Jerennian.

He looked tousled and rough. I recognized someone who was fresh out of hibernation, though I didn't think through the implications of that at the time. Ste Etienne and I were two fifths of the way into a gladhanding reunion before Advent cleared his throat.

Jerennian looked at us, stony-faced.

'For the record,' said Umbar crisply, 'Debrief of liaison Ceelander J and engineer Ste Etienne M, following their recovery from Prospector413b-parenthesis-one, their reports having been taken into consideration.' She made a note on her terminal, probably linking our report files with the recording. 'Present, Chief Director, Asset Acquisition Director, Technical Oversight, plus technician Jerennian M present for his perspective as the only other surviving member of the Prospector413b Special Projects Team.' It was all rattled off with admirable brevity.

Advent shifted, waiting to be sure he wasn't going to tread on her heels, then nodded. 'We've listened to your reports. Under other circumstances, I'd say they range from unhelpful to unbelievable. But . . .' A very grudging 'but', and I had a sudden jolt of wondering what had happened. What contact had there been, with the Shrouded? Except it was nothing so dramatic, and he went on with, '. . . the data from your vessel confirms most of the details. Even the more bizarre ones.' Of course, it hadn't just been our frail human bodies recovered from the pod. All the compressed data, visual and electromagnetic both, and probably even our increasingly deranged conversations, had all theoretically been recorded for retrieval. I wasn't actually sure what'd been retained, and what had been overwritten, because the pod surely didn't have the storage capacity necessary for all that wealth of information. Apparently enough had been preserved for us not to look like delusional maniacs.

'Obviously,' said Advent heavily, 'this is not the debriefing I'd anticipated, with regard to the Special Projects team.' He looked unhappy, although it was mostly the unhappiness of a man with extra paperwork to do. 'Under ideal circumstances, a one-to-one with Director Bartokh would have been the preferred outcome.' He looked at the two of us – the three of us, really, because Jerennian was definitely not on the management side of things. I wondered if we were somehow expected to have preserved Bartokh, and how exactly they thought we could have managed that. I didn't feel there was some obvious opportunity we'd missed, down there, not with everything that had been going on.

'Your reports are filed,' Umbar added, with the rider of *such as they are* hanging unspoken in the air. 'Do you have anything to add?'

Of the two of us, Ste Etienne and me, I was the talker. The social animal who couldn't abide an awkward silence. Except that quirk had rather rubbed away, given the long quiet spaces the pair of us had endured down on Shroud. You'd have expected me to be running my mouth off the moment I was given an opportunity. A wellspring of commentary, suggestion and speculation. But I didn't. Because, of the two of us, I was also the one more generally capable of reading the room. And the room didn't read right, frankly. This was not the debrief I had expected. The fanfare and the parades were pointedly absent. I wasn't sure what we'd done wrong, exactly, but this was the sense I was getting.

Ste Etienne hadn't picked up on it, though. She did have

something to add. 'We're ready to work on it,' she said bluntly. 'Or, I can't speak for Juna, but I am.'

Umbar exchanged the smallest sliver of a look with Advent. 'Clarify,' she invited, with the air of someone giving you the opportunity to inspect the big hole in the floor, while they stood innocently behind you.

'The Shrouded,' Ste Etienne said. 'Or – the natives. The species. Whatever. Unlike anything we've ever met before. And we must only have seen the very edge of what they can do. You'll need someone working on them. We've got the experience. A new Special Projects.'

Another brief sidelong look, as though they'd been allocated a single exchange of glances for the whole meeting and were rationing it out. 'Is that,' Advent asked pleasantly, 'your recommendation?'

I wanted to kick Ste Etienne. If we'd been meeting old-style, with an actual table, then I could have done. As it was, we were just strapped to the wall and everyone would have seen.

'Sure,' she said. 'I mean, obviously. This is something completely new. We need a team.'

Advent looked embarrassed, frankly. He grimaced a bit, then practically nudged Umbar into taking over. She scowled at having to say what she plainly felt should have been obvious.

'Given that fifty per cent of the current Special Projects team have been killed as the result of an industrial accident, it is not the considered opinion of Opportunities that a new team should be formed. Nor even that the current team should persist.'

Ste Etienne blinked. 'What?'

'This is why I'd rather have had this discussion with Bartokh,' Advent grumbled. Meaning it would have been Bartokh breaking the bad news to us, not the great man in person. 'Special Projects is a failure. It accomplished nothing, and consumed an unconscionable amount of general resources. It was not a profitable endeavour.'

'But what happened,' Ste Etienne ploughed on, 'in orbit. The accident. That wasn't *us*. We didn't have anything to do with that.' Before anyone could handwave that away, she went on, 'Look, you've seen what we ran into down there. What we survived. What we discovered. The ecosystem, an actual intelligent species. Tool-using. *Inventing*. We have to study that, right? Juna and me, we're your experts.'

Advent again looked embarrassed for her. Jerennian was shooting her looks to try to get her to shut up. I just stared at the wall.

'For the avoidance of doubt, we accept that Special Projects did not contribute directly to the incident which killed Skien H and Rastomaier S, and indirectly led to the death of Director Bartokh,' Umbar said. 'However, even before that, the resource deficit represented by Special Projects had become a cause for concern. Your team achieved nothing, I'm afraid. An error of judgement that Opportunities can only take as a learning exercise for the future.'

'What do you mean, nothing?' Ste Etienne demanded. 'Terwhin, you've *seen* what we went through. That's not nothing! It's an incredible opportunity—'

'Speaking for *Opportunities*,' Advent broke in, 'we do not think so. Since the termination of Special Projects, the harvesting operation has begun, and a return is at last being realized from the moon.' He nodded to the sour man on his right, who plainly felt he didn't need to be there. 'Orbital infrastructure has been re-established and is being expanded as we speak. We have a viable operation here now. I don't know where you feel this research effort you're proposing *fits* into that structure, but I don't see a place for it. I think we can skip to final assessment?'

Umbar nodded. 'Very good. To finalize the accident inquest, I think we can confirm that Special Projects did not materially contribute to its cause or outcome, only to the material lost as a result. If Director Bartokh was here, then I feel we'd probably note that as a demerit on his permanent record. But as matters stand, it is the decision of Opportunities that no direct blame accrue to the wage accounts of the surviving personnel.'

Jerennian relaxed. He was a big enough man that this was visible. I hadn't even realized that was why we were here. To determine whether the sunk cost of Special Projects was our fault, and would screw over our futures, rather than just being buried with Bartokh.

Ste Etienne made a noise. Her eyes were very wide, staring around as though what she was seeing was as strange as anything we'd been presented with down on Shroud.

'To finalize Special Projects,' Umbar went on, 'I think we can formally discontinue the team, effective as of now, and chalk the costs up to experience. This leaves Acquisitions as the sole active effort relating to this moon, and we can follow up with them on the next review cycle.'

'Agreed,' Advent said. It was as though we weren't there any more.

But Ste Etienne didn't consent to not being there. 'So what's next for us?' she asked Umbar directly, the woman she'd been cultivating, *buffing*. 'I mean, we've all three got the expertise, right? And nobody else has seen or done what Juna and I have.' *We're heroes, right?* Except I had already seen that particular ship depart. Sailing away, catching fire and sinking with all hands, frankly.

'I regret to say,' Umbar told her primly, 'that we do not currently require your expertise. There is no active role for you within our live operations. And, I'd note, existing teams might find your presence disquieting, given what happened to your last assignment. You will await future deployment in the fullness of time.'

In the fullness of time. It was the standard phrase. The one you never wanted applied to you. It meant they had no use for us, any of the three of us. There were other technicians and other engineers on the *Garveneer*'s roster, and most certainly other people who could do whatever the hell anyone needed *me* for. We were bad luck. Associated with three deaths and a lot of wasted resources. We were Jonahs. Nobody would ever write that kind of superstition into our records, but the stink of it would follow us anyway. So it was back into hibernation for us, and we'd be right at the end of the queue to ever be woken up again.

As we were unstrapping to leave, the sour man finally found some words. They were muttered, and for Advent's ears rather than mine. Wrapped up in my own woes, I didn't really process them at the time.

'Weren't you going to ask, Chief Director,' he said in a low, gravelly voice, 'about the lost remotes. The damage?'

For a moment Advent looked like he was going to call us all back into session, but then he plainly decided it wasn't worth it. We'd taken up more than enough of his valuable time. Damage happened. Remotes were lost. Conditions on Shroud were adverse, after all. The asset-stripping operation was expanding despite it all, and that was the important thing.

Hibernation. I realize it's a logical system – the only way, really. For the sort of operations ships like the *Garveneer* are meant for, it swings from needing all hands on deck to only requiring a skeleton crew to watch the colour of the lights. In those quiet times, which can last for years, it's not resource-efficient to feed the whole crew or give them air to breathe, entertainment, or suffer the wear and tear on quarters, let alone stump up their wage-worth. I mean, you can't expect to sit around and be paid just because they'll need you at *some* point.

It's a good system. It is. It's necessary.

When you're facing the beds, though, it doesn't feel good. To know you're surplus to requirements and they're going to shelve you until further notice. Like everyone, I've been there plenty of times, and you don't get used to it. It's like death, to some atavistic part of the mind.

Shelved. That's the word people use. And it happens in a place called the Long Dorms. An octagonal-section shaft, with one face used as a floor, the other seven lined with slots. You go into a slot and they shoot you with the sort of drugs the pod was never equipped with. And you don't dream.

I looked at Ste Etienne. She looked at me. Jerennian didn't look at either of us. They'd brought him out of this just for the debrief, the formal closing down of Special Projects, and now they were shelving him again.

They hadn't thrown him away immediately, he'd said. After

the accident, Special Projects had limped along. Meaning him. Umbar had taken a look at what we'd achieved, what was left of it. There had been talk of convening a new team to pick up the work. But the loss of resources and personnel had killed that off in the end. They'd made the call to just start asset stripping. The unique aspects of Shroud were judged not to be enough of either a problem or an opportunity to expend more time and materiel on. And so Jerennian had been shelved.

Ste Etienne clasped my arm. Her lips were pressed tightly together. Her look was bitter, but at least the bitterness wasn't directed at me. I couldn't have borne it if she'd ended up thinking this was my fault. Without that, at least we had each other, for the moment. Obviously, there was no guarantee we'd be woken up again at the same time. Our skillsets were very different, and it'd been pure chance that threw us together. We might never actually see each other again. One of us might be woken far into the future around some different star, to find that the other was already old and retired to a farm station, like the dream was. Or else dead. Or just unfindable, fate unknown. Because information wasn't free in the Concern, and wasn't evenly distributed across human interstellar endeavours. Sometimes you just couldn't find something out.

Nobody stamped their foot and complained. Nobody rebelled. We just took three seconds longer than regulation required, sharing that look, and then climbed into our claustrophobic tubes. Finally, they shut us down.

6.2 DARKNESS

There are useless thoughts – ones that go nowhere. These cannot even be projected from segment to segment within the quiet of a nest, let alone to any useful distance outside. It is not the competition of the world that strangles these thoughts. It is the very nature of the world itself.

I experiment with my new acquisitions from the Stranger's associates, both the material I appropriated, and the replica conduits I have built. The useless thoughts, which I can generate from the crystals in my bodies, but are too feeble to survive outside them, travel very differently within the minute hollow spaces of this material. They fly. Converted to such a medium, the pattern of my thoughts moves from one terminus of the tendril to the other more swiftly than anything I have ever encountered. Faster than thought itself, it seems. I burn resources and accelerate the rate at which I process information, yet I cannot match the swiftness of those weak thoughts when they are passing through the hollowness of the tendrils.

Gathered in my home, I consider my map chamber and the separate globes, each one an island in the world where thought exists. Where an *I* which is cut off from *me* projects thoughts that touch the world. Each one alone, surrounded by the unknowable. But what if that was not so?

I can manufacture more of these hollow tendrils as easily as I can the spars of a body. The stuff they are made out of is plentiful: I can grind it up from the rocks and then extrude it in its pure form, sucking it empty of air as I do, so the weak thoughts can travel more easily down it.

I proceed to form great lengths of it and think about the colony by the sea, where the hot chimneys are. Not so far away, as mere distance goes. Very far, though, due to the hostile country between us, with the ghosts beneath the mountain and the ice-spitters. It has always been hard to exchange thoughts and ambassadors with my nearest neighbour.

Using the airship design that Otherlike mind gifted me, I can carry an unbroken length of tendril from my home to the coast. My thoughts will then travel along it, swifter than the storm winds. Swift enough that, when they reach that Otherlike self, they will *still be my thoughts*, and the thoughts which return to me will also be mine. Two separate Otherlike homes will become one. *We* will be *me* once more.

Even as I think this, I am putting the experiment into practice, and reach out to my Otherlike there. With all my resources bent to the effort, and my multifaceted perspective examining the problem from all sides, the task is achieved quickly and efficiently.

We connect. Thought travels clearly, untroubled by the tumult of the world.

A cascade of memories meet and form a new whole, and my mind expands. I understand a whole history of separation and, rather than these details mingling and dissolving amidst the wealth of everything I already know, I have the resources to hold it all in mind. These parallel paths I have taken now rejoin as one.

I become greater, and in the echo of that I know only a hunger for more growth.

I consider my maps.

6.3 LIGHT

You don't know what you'll see when they wake you. Or how much time will have passed. In the habitat tanks, growing up, it was hard to make firm friendships because everyone was in competition for the job which would take you out of that place and into space, where you'd have a chance of something better. Then when you were in space, you could only ever make friendships of convenience, based on who was around you at any given moment. It was why someone like me was genuinely useful. Social glue, instant facilitator, and centre of cohesion for any ad hoc group of people. I knew they'd need to hoick me out eventually, and I'd start the whole business again. Getting to know people. Making myself useful.

I expected another star system, really. The *Garveneer* having crossed the vast gulf of space, leaving the mined-out husk of Shroud behind it. Or else I expected Shroud, alongside a handful of other moons and planets, bristling with orbital infrastructure, heavily exploited and now being converted to waystations and supply farms, or other permanent stopovers on the road of human destiny. When they unshelve you, you have no sense of how much time has passed, a day or a decade, and the process interferes with the human brain. Or rather they have to shut down the part of the brain which counts the hours, because otherwise it messes with you to the point of psychosis.

I should probably claim I didn't have any idea about my situation until they actually started briefing me, but that's not true. Ste Etienne was there, similarly deshelved and struggling into her crew overalls just like I was. And Jerennian too, looking hunted, because going in and out of hibernation in quick succession isn't good for your metabolic balance and makes you feel strung out. There was absolutely zero reason for the three of us to be out at the same time, unless something had happened on Shroud. Something soon after they'd put us under.

We received our automated marching orders, which said we were booked for another face-to-face with Opportunities, but first we had dossiers to read, to bring us up to speed. This communication gave us a working time frame. We'd been under for only forty-one days.

They didn't split us up – that was the next clue. Normally it would be each to their own solitary cubicle to absorb the brief. Keeping us together meant we were nominally a team again. Fifty per cent of the old Special Projects, brought back for one last tour. We ended up sat down in a triangular room, what they called waste-space. It was something that should probably just have been a bulkhead but had been hollowed out, so three deadbeats like us could anchor ourselves to the walls and drink stims, as we absorbed what had been going on. Well, Ste Etienne and I were doing that, while Jerennian hacked his terminal and searched for us in the *Garveneer* active crew hierarchy. We were assigned to a team called Remedial Work, which was simultaneously vague and unflattering. It was under the jurisdiction of none other than Technical Oversight Terwhin Umbar, and there was also someone called Ducas FenJuan on the list, who was down as an exobiologist, like Rastomaier had been. Their name and face nagged at my memory, until I recalled them as the person who'd been about to take over from Bartokh right at the outset, until they weren't. The one too enthusiastic about aliens, and who'd been abruptly shelved when that turned out not to be what Opportunities wanted after all.

This all told us what we needed to know, more than the briefs, really. It told us that Shroud had gotten ornery.

As Jerennian had been otherwise engaged, we summarized the briefs for him before our date with destiny and Operations.

'Basically,' I said, 'things have started to fall apart.'

'They do that,' he agreed.

'Acquisitions is just getting up to speed, moving out from the pole,' Ste Etienne expanded. 'Only they've started losing stuff. And the data's muddy, right? Everything's either connected directly to the anchor point, or it's autonomous and can't check in because of the local conditions. Or global conditions, pan-lunar conditions, whatever. They've had the regular amount of breakdowns. Terrain, mechanical failure due to pressure or gravity, and, because they're mass-producing on the cheap, alkaline corrosion. All within tolerance, just about what you'd expect.' There was no point building everything to last forever. You had to work within a band that said the number of breakdowns you'd get would be outmatched by the resources you were saving by cutting corners in the manufacture. This was what efficiency meant.

'But now the failure rate is going up,' I noted, 'and a lot of the time, when they try to recover failed units, they can't locate them. Which, again, might be local conditions, given what it's like down there.' I shuddered, without warning. A little PTSD leftover from our ordeal, and why not? Ste Etienne and I were the only humans in the universe who had actually experienced *what it was like down there.*

'Remote or wired losses?' Jerennian asked.

'Both,' Ste Etienne said, brisk and businesslike, no shudder from her. Or none that she would admit to. I reckoned it was there, though, now I actually knew her. 'Severed cables. Units just disappeared without signal. And sometimes the recovery units don't come back either.' She made a weird sound. 'I mean, in the dark, without any kind of long-range scan capability, what are you going to do, right? Thing you're looking for could be just over there, in arm's reach almost. Or just past that. Or just past that. If it's gone dead and can't even signal, you'd never know.'

'But they think it's the aliens,' Jerennian said. Thinking of the surprise exobiologist added to the list. As well as including *me*. Maybe they wanted Ste Etienne and Jerennian for their experience in making our tech work with Shroud's conditions, but right then *I* had a very specific and narrow competence that might qualify me as more than just social glue.

'There have been sightings,' I confirmed. 'Some that came right up to the anchor. Others they ran into further out. They've found part-disassembled asset units too. They've . . .' I grimaced.

'They shot them the fuck up, is what they did,' Ste Etienne filled in. 'A bunch of the locals turned up near the anchor and started taking some of the processing plant there apart, so the defence drones kicked into gear. Fair enough. What're you going to do, I guess? Might just have declared war, though.'

'War's a human thing,' I said, but hollowly. Maybe war was actually the one universal thing, after thermodynamics. Everything fights, in the end. The aliens probably fought one another a lot, all those scattered polities of them. We were just one more presence on Shroud to be squabbled with. They weren't going to know what hit them.

'So they think, what, it's retaliatory action?' Jerennian asked.

'They don't know what to think. Maybe it's just aliens doing alien shit,' Ste Etienne said. 'They want our take, anyway. We're back on the mission. "Remedial Work". Sounds like the team for slow learners, doesn't it?'

I thought about the Shrouded, who had proved themselves anything but.

We didn't have the pleasure of Chief Director Advent's company overseeing the meeting. It wasn't that important. I think he'd only been there to shut down Special Projects because someone as high-worth as Bartokh had been killed, and he wanted it clear that he had his finger on the button. But Remedial Work was, as the name suggested, not a major star in the *Garveneer*'s firmament. Instead, we got Terwhin Umbar, who looked harassed, and whose own celestial body had plainly fallen a few rungs. There was a different surly man, the new head of Acquisitions, who also went entirely un-named and had nothing to input. Ducas FenJuan, who was just as neat, small and outwardly genderless as I remembered, turned out to have become quite a fan since our original brief meeting.

'I wish to say,' they told Ste Etienne and me, 'it's an honour. What you went through. Remarkable. The data you retrieved.' Then Umbar scowled at them and they shut down, staring at their hands.

Umbar ran through the usual meeting opener, for the record, and then glared at all and sundry.

'This team is convened to work out if the planetside inefficiencies we're experiencing result from activity by the local fauna,' she said. 'And to present solutions to Opportunities for consideration. If you can provide reasonable cause and remedy, you'll be given a construction budget, and the team's lifespan will be extended to cover manufacture and deployment of whatever your solution is.' Meaning we'd be given longer off the shelf, and earning our worth. More chance to impress and escape the resource sink that Special Projects had become, through no fault of its own.

'I . . .' Under her glower, I licked my lips. 'Who's Director of the team?' Of course, we hadn't officially seen the organization chart Jerennian had tracked down, but there hadn't been any name in place for the post. I was thinking it might be me, and I wasn't sure how I felt about that.

'I am,' Umbar said, as though the words came with thorns. That didn't make sense, though. She was head of Technical Oversight for the whole *Garveneer* operation. She was important. And everything we'd been shown about Remedial Work insisted it was trivial. Probably a wild goose chase, and it was all just mechanical failure and accidental damage the whole way down. No need to suspect anything . . . unusual going on.

I felt something inside me, halfway between a thrill and a knife-twist. Something unusual *was* going on.

'I'm devolving day-to-day operations to FenJuan as acting deputy, who has been studying the data you recovered from planetside Shroud,' Umbar said. 'They'll want to talk over some points with you, and workshop some possibilities. *I* want you to find a working explanation for what we're encountering planetside, and a fix for it soon after that. Do not, therefore, spend too long on empty theory. We're well into the resource phase of our operation here, and we don't have time for pure science.'

FenJuan nodded minutely, and then Umbar kicked off out of the room, the silent Acquisitions man following after her.

They had both seemed stressed, I sensed. Under pressure. A pressure they were diligently trying to export down the chain to us.

After they were gone, FenJuan opened up what felt like about a hundred screens, all showing either raw data or image clips from our trek across Shroud. The Shrouded themselves featured heavily in the latter, needless to say, and much of the former comprised EM signal logs. Not all from our journey, I noticed. Some looked like they'd been scraped from Acquisitions robots, or their attendant infrastructure units.

FenJuan wasn't one for small talk, but there was a reason that sort of thing was part of my own specialized skillset. Between performance pressure and the constant shifting of company from waking to waking, nobody was really encouraged to sit around chewing the fat when they could be working, or talking about work. Your time was the Concern's time, as they said. Hence I had my mouth open, ready to hand around introductions and pleasantries like canapes, when Ste Etienne just bulled right in.

'You think what, then?' she asked FenJuan. 'They did it?'

The exobiologist's crisp little smile didn't shift. It was, I thought, their default expression. Their defence against rude people like Ste Etienne constantly trampling over them. 'The matter is still under consideration—' they tried, but the engineer wasn't having it.

'You've got a theory, though. I mean, I know what *I* think.'

FenJuan lifted an eyebrow. Everything they did was very precise, like someone I knew who'd been badly burned once. Though they'd been fixed up and weren't hurting any more, the memory of it was like a brake on each movement they made.

'You think that it is the . . . Shrouded, Ceelander calls them in her report.'

'Even from the dossier and what just got said,' Ste Etienne agreed flatly. 'Look, you said, about what we went through, so you know. You've studied everything they could pull out of the pod.'

FenJuan nodded.

'So they take stuff apart. And they work out how it works. You reckon we go much further we'll run into their own resource drilling rigs? What do you think?'

'What Mai means,' I put in, before things could get too adversarial, 'is that they're obviously a complex species, capable of tool use and innovation, and curious about their world, and about us. Very curious about us, to the extent that various groups of them actually followed along. A lot of them actually died protecting us, or just because we went into places that were dangerous for them.'

FenJuan nodded.

'And now we've killed a bunch of them ourselves. I don't *know* what they were thinking before, about us, or what was really behind all the stuff they did, but it's surely changed after we've shot a load of them.'

FenJuan nodded once more, the same motion repeated exactly. 'It is notable that they haven't approached the anchor point again after that incident. Sufficient of them survived the encounter to communicate the danger to the local colony, at least.'

Something fell into place, in my head. I glanced at Ste Etienne and met her looking right back at me. We knew something our betters hadn't worked out.

'I have been studying the interactions of the . . . Shrouded.' FenJuan plainly didn't like the name. 'Not their interactions with you, but amongst themselves. I am . . .' They actually leant forward and lowered their voice, as though there was any chance of avoiding the *Garveneer*'s ubiquitous systems overhearing us. 'I am constructing a theory regarding the Shrouded. Just a theory. I . . . have been in trouble for my theories before. I was only taken off the shelf because a certain level of free speculation seemed called for in this case.' Their hard little smile, like the faceplate of a helmet. 'Many of them died. For you. The Shrouded.'

'I mean, we don't know it was for us,' Ste Etienne said warily.

'Here.' The screens showed the tubeworm creature, erupting from its hole and grabbing the pod. 'They do not want to go near, then you are attacked and they commit themselves. Many die.'

'I thought perhaps we'd somehow established ourselves as one of them,' I said. 'So they instinctively defended us.'

'They die. You compare them to ants. When you saw their farm. I agree. They are social animals, like ants. Or more than that. The signal.' They rapped on the wall, three evenly spaced beats. 'There is something strange about it. It is passed between them, between colonies. Different groups recognize you, and fall into the same patterns of behaviour. A transmission of knowledge? Language and learning? A culture?'

'Are you asking or telling us, or what are we doing exactly?' Ste Etienne queried. Jerennian snorted.

'I am concerned.' FenJuan's lips pressed tight, bloodless. 'I am concerned about what is going on, on Shroud. I am concerned that we will reach conclusions not accepted by or useful to Opportunities.' And it sounded like they'd been shunted back on the shelf more than once on account of that odd theory-making of theirs. Probably they saw Remedial as their last chance to get back into proper circulation. 'Their electromagnetic interactions with the world are very complex, as recorded by you. They still respond to your signal. I programmed a drone to send it, and several groups of Shrouded responded. And dismantled the drone, in one case. Also there are the cables . . .'

'Yeah,' Ste Etienne said slowly. 'I saw that.'

I hadn't, so they had to find the footage for me. A group of Shrouded, the basketwork tardigrade shapes with offset legs and multitool heads, advanced into the light of a drilling installation, a light they wouldn't have been able to register. At the back was a bigger one, the same general shape but twice the size of the others and bulked out by . . .

Jerennian barked out a single obscenity, and I could only

echo it. A drum. A revolving drum set into the thing's exoskeleton. It was just rotating on pivots there, as though it was no great achievement to not only reinvent the wheel, but make it a part of their bodies. And on that wheel was what looked very like a spool of cable. The whole being, I thought, was a weird echo of the cable-laying robot we'd grabbed to send a signal home.

'They playing dress-up now?' Jerennian asked, frowning thunderously. 'What the hell is going on?'

'Officially, maybe nothing,' said FenJuan primly. 'In my thoughts, absolutely something. The behaviour of the Shrouded is adjusting very swiftly to our presence. As you discovered with their incorporation of airship technology.' Incorporation, not adoption. Because technology wasn't something the Shrouded *made*, but something they *were.*

Ste Etienne and I exchanged that look again. Time to play the hole card.

'You said,' I observed, 'the local colony.'

'Current theory is that our activities are impinging onto the territory of one of these Shrouded nests, in the vicinity of the resource-stripping operation.' FenJuan's eyes were fixed on me.

'Except that region is . . . desert. There's no life we saw, outside of some volcanic oases,' I said, wondering how many of those tiny islands of life, each one unique perhaps, had been obliterated in our operations. 'We didn't see our escorts predating on those at all. They kept well clear of them. So . . . I don't think there necessarily *is* a local colony. That's not what we're running into.'

FenJuan's face remained carefully composed. I decided that nothing I'd just said came as a surprise to them, but it ran against what Opportunities pronounced was going on, and therefore had been a step they'd not been willing to take themselves.

'Kindly continue,' they said, on the basis that if there was a stupid thing that needed to be said, it was always better for someone else to put it into words.

And saying those stupid things was, sometimes, what I was here for. 'If we're up against something,' I said, 'it's something bigger. More organized. You're right, about their interactions, and the way information seems to be passed between them. I don't think we actually understand the scale of what we're dealing with here.'

6.4 DARKNESS

I connect. That is my obsession. The pace of interconnection increases, and so do I. The more I do, the more I *can* do. Each cluster of Otherlike that rejoins me brings with it not only local knowledge, but a scatter of novel innovations and techniques. Things that a moment before I had never thought of, but a moment later have always been a part of my life since I invented them. Each connection expands my ability to view the world, with the multiplying of perspective, and the reach of my thoughts as they explore my environment. The sheer, brute power of my mind to analyse what I discover, understand it and put it to use.

I feel as though I am racing through time. Would I ever have reached this level of interconnectivity across the world without stumbling across these new ways of doing things, these alien tools? And they are alien, I understand. They have been manufactured by a process quite unlike anything that occurs anywhere in my world. I have received impressions back from the great vertical spire erected in the heart of the ice desert. It goes up higher than my thoughts can follow it. I know, now, that it was not built from the ground, but from the sky.

As there is now enough of me, I have had thoughts about the ground, the sky, and the cycling of signals from beyond, and how they relate to one another. I am still ordering these thoughts, but I feel a web of interconnectivity there, for me to send my thoughts along too. The relationships of time and distance, speed and mass. The complex ways one might describe, comprehend and manipulate such things.

That I have chanced across an unparalleled way of connecting two points, two communities, two parts of the same mind, does not mean there is not an even better way permitted by the functioning of the world.

I have also become aware the Stranger left a legacy. Something is happening in the desert, spreading from that structure in the centre outwards. Another interconnected network, radiating the sort of broken, mindless thoughts the Stranger was wont to give vent to. A growing encrustation of alien life.

I investigate. I send aerial bodies and use them to land exploratory parts of myself to inspect and report on it. Some return only piecemeal, but more return with bounty. What I would previously have disassembled into gross mechanical pieces, and thought myself learned for doing so, I can now appreciate at a variety of different scales. Once I cracked open the Stranger's other segment, thinking it was an ambassador with a message for me. I learned only disappointment from that, because the *I* which acted was not wise enough to understand what I possessed. Now my wisdom extends across everywhere my maps once recorded – those maps which are so much more detailed now and reflect the entire surface of the world to me. I can craft the smallest, most subtle thoughts to explore the fine detail of the alien segments I take. And so I follow the myriad connections within them, explore their logic, generate energy to activate this part or that, and watch it shuffle through its routines as though it were still whole and alive.

Each experiment teaches me new rules and patterns. Each rule and pattern becomes a tool for understanding the context of what I have uncovered. And all the while your alien presence spreads. I grasp your nature now. You hunt and harvest, just as I do. And now I have your works, I understand why you hunt, and the uses to which the harvest will be put. Like all things, you seek to build more of yourself. I find no fault in that, save that your progress, accelerating as it does, must needs negate my own.

I accelerate too. Exponentially. Not long ago, I explored your capability to destroy my segments, even as I was exploring my capability to destroy yours. Now I consider a different avenue of attack. Your segments and your structures are constantly leaking their thoughts into the world, after all. Where thoughts escape, so might they creep in. I have studied the logic of your bodies. I have the breadth of perspective to understand how they are made. The signals that carry meaning back and forth within them are not so very different to my thoughts.

I cannot infiltrate you with my bodies. Perhaps I can do so with my mind.

6.5 LIGHT

'The problem is,' I said, 'they don't know what they want us to do.' *Us* meaning Remedial Works.

'They want us to fix it by magic,' Ste Etienne said flatly. Meaning, find a remedy, as per our team designation. What remedy, though? That was the problem.

There had been a handful of other speculative approaches to the anchor by the Shrouded. We'd seen the images and data – most of the actual fighting had been at a range beyond the installation's lamps, as we could use other wavelengths to target with, just like the locals. The Shrouded had turned up, been shot at and retreated. There had been no mad storming of the guns by a horde of clattering alien monsters. No panic, either. It was like they were making an offering of blood to us. Except, after forcing myself to watch again, over and over, I reckoned they were feeling out where our boundaries were. How close they could come before they provoked a violent response.

And this knowledge had, apparently, sunk in. They weren't going anywhere near our central hub at the pole. But the resourcing operation was spread out across an enormous area by now. We were installing automatic guns at the other major hubs, but isolated robots and even whole exploratory mines were just being stripped away. It didn't feel like a concerted assault so much as opportunism. That 'feel' was laden with our human expectations, though.

Ste Etienne and I were in one cubicle, studying what we had and trying to build models of what was going on. So far, our progress had been slow. Not only because we had so little to go on, but because our allocation of data resources was pitifully meagre. Every analysis we set our algorithms onto took longer than the last, as the other departments throttled us. It's never any fun being at the bottom of the pile.

Things were perhaps also slow because by the time we had a model of what was going on, what was *going on* had accelerated and changed to something else. The Shrouded were never still. We remembered that much.

FenJuan was in the next cubicle. Being Acting Deputy Assistant Director of Remedial Works, they got one to themself. It wasn't exactly the lap of luxury. They'd had Jerennian rig up a storage block so they could play and replay the recordings of our trek, watching the Shrouded bustle about in the circle of light around the pod. Occasionally they'd make a note.

Jerennian himself was in the big workbay that now occupied the hollowed-out belly of the *Garveneer*. All sorts of experimental stuff was being built there, up in orbit where there was no chance of sudden alien incursion, and where moving heavy items about was nominally easier. Jerennian was our liaison with the personnel and resource-heavy team known as Safeguarding, who were theoretically constructing equipment based on our recommendations. Since we had no recommendations yet, they were just throwing things at the wall, designing stuff that felt like it might be useful.

Midway through one day, he called in remotely to explain what was going on, but the connection stuttered, sliced into separate frames, with his mouth never moving, just jerking from one frozen moment to another. His voice came over with a robotic judder.

'Are you having these data outages?' he demanded, or I think that's what he was saying.

'Wait, you are too?' Ste Etienne demanded. Because we'd been grumbling along, like I said, on the assumption we were just at the back of the queue for resources.

'Everyone is.' Jerennian's voice twisted and warped. 'System's on a go-slow. Look, they are going nuts here. You know how it is. Physical resource budget through the roof and nothing to spend it on. You're not going to like what they're working on.'

He turned out to be right about that one, but we'd had to make him repeat it three times before we could reconstruct the sentence, so our focus was on the connection, not the actual problem he was talking about.

The image of Jerennian froze in the middle of him rubbing at the back of his neck and scowling. His voice came through quite clearly. 'It's like it's an attack.'

After the fruitless call had terminated, Ste Etienne and I looked at each other.

'How can it be an attack?' I asked blankly.

'It can't,' she said. 'Get back to work. Mike was always nuts.'

'I thought you liked him.'

'Sometimes you like nuts.' She shrugged, and then we both realized FenJuan was standing behind us, at the open side of the cubicle, with the datastore under their arm.

'I will run something by you,' they said. There wasn't really room for three people and a datastore in there. There wasn't really room for two without one. They came in anyway, because they were direct like that. Practically sitting on my lap as they brought up their screens.

'The latest images. From our roving drones,' they said. We saw trains of Shrouded crossing open country, caught briefly in the drones' lamps, and then again as the remotes circled back to take another look. They had one of the big cable-layer aliens there. In another image there were no aliens, but a line of disturbed earth, very straight. A third set of images showed an ascending spotlight down a twisted spire, almost entirely covered with veiny vines and great flowerlike extrusions that opened and closed, raking the air with feathery fronds. Shrouded were coming and going at its base, and I thought the spire itself – as the drone's view raked up it – might have been something like the gigantic pylon we'd come across, before it became overgrown.

'A nest. We think,' FenJuan said. It might have been, because no two Shrouded nests we'd seen had been the same.

They paused the image, closed in on one quarter. It was hard to make out, given the way Shroud chewed away at image quality, but we saw a bundle of serpentine tendrils leading inside.

'I'm going to say something,' FenJuan said. 'It is the sort of idea I used to have. That got me shelved. But if they put me back in action, it's the sort of idea they want from me, I'd guess.' A shrug. 'The Shrouded have a suite of electro-magnetic senses. It is part of how they navigate their world. And it is part of how they interact with one another. You discovered this, because to a very limited degree you were able to insert yourselves into that conversation.' When they were talking science, their whole manner changed, a hidden enthusiasm stepping forward that had been lurking away in the darkness until then. 'Hence they exist in an electromagnetic world. Much like our technology does.'

They waited for us to shout them down, but we'd seen what we'd seen, down there. The proposal sounded entirely reasonable.

'You also established that they are possessed of an acquisitive intelligence. Capable of recreating and adapting designs that they are shown. They had not taken to the air naturally, even though they were possessed of the tools to do so. When the feasibility of the idea was demonstrated to them, along with a viable design, they were able to adapt the original to their needs almost immediately.'

The scent of where this was going lay heavy in the air. Ste Etienne and I exchanged looks.

'Yeah, well, mechanical, that's one thing, right,' the engineer said, almost nervously. As though some inquisition was about to have her reshelved for heresy. 'But you're talking . . .'

'They live in an electromagnetic world,' FenJuan said. 'Who's to say that the functioning of a complex electronic device is not as apparent to them as a physical one? It would exist in their sensorium as an informative object, far more than it would to us.'

'Who's to say this, who's to say that?' Ste Etienne tried for dismissive and couldn't quite get there. 'You're saying they're using the cables, and making their own cables, just to holler down them like two kids with a length of hose?'

'If they can produce a signal then they could make use of the infrastructure we have demonstrated to them,' FenJuan went on implacably.

'Well I don't know if you're up on fibre-optics but they use light. And light's the thing they don't have down there.'

'Light is electromagnetic radiation.' FenJuan shrugged. 'Not a naturally useful wavelength, given the conditions on Shroud. You're right. Nothing seems to be generating it. But if the electromagnetic spectrum is so integral a part of their being, and if they are as capable of analysing their world as they appear to be, then they would surely register the band where visible light wavelengths exist. And understand when the use of that band is demonstrated.'

I watched Ste Etienne open her mouth three times with put-downs she plainly couldn't quite get behind, and then I said, 'To what end, though?'

For a moment FenJuan's face closed up, but they obviously decided they'd gone too far anyway. If their subordinates were going to shop them to Opportunities, they'd given us enough ammunition now.

'Connectivity,' they said. 'I have a lot of circumstantial observations to suggest that the Shrouded act as a hive mind, especially the way they're willing to sacrifice individual organisms like they have against the guns. You say ants. I think they are like very clever ants. Communicating not through scent and touch but by electromagnetic wave. And now, thanks to us, communicating between communities using our technology. I think we have given them the ability to organize at a grander scale than nature would have permitted. Organize against us.'

We exchanged looks. If nothing else, it meant we had something we could report to Opportunities.

* * *

We tried to report remotely, but in the end Umbar – Technical Oversight I'll remind you – got fed up with the poor connections. It was a plague, it seemed. Every system process was dragging its feet, and every data channel was glitchy. A whole new culture of actually interacting directly with other human beings outside your own team had sprung up across the *Garveneer*. I rather enjoyed it, honestly. We'd have awkward ambassadors come over from Safeguarding or Resource Acquisitions or Infrastructure Oversight, new faces we'd never, under other circumstances, have met. Or else I – as our designated people person – would have to navigate the convoluted spaces of the ship to find other teams and talk to them about resource sharing, and ask where that data analysis was we were after. Sometimes there would even be treats – a scavenged or printed delicacy, or some bootleg stim drink they'd banjaxed the printers for. In those days, I met more different people than I had for subjective years before. Maybe more than any time since leaving the crowded confines of the habitat tanks.

But it was the whole team who had to turn up before Umbar. If she was going to get her feet dirty with the hoi polloi, then she wasn't going to make the effort for just me. She looked increasingly bad tempered every time I saw her, honestly. I reckoned Advent was riding her hard. On this choice of words I couldn't help thinking of her – still, no matter how long had passed – being *buffed* by Ste Etienne, or of Ste Etienne being *resourced* by her. Umbar, uptight and constantly bearer of the bad news that was resource allocation or assignment details, with her human side jealously locked away. Save that at least Ste Etienne could bear witness she had one.

We'd tendered a written team report, but FenJuan had to summarize anyway. It was an extraordinary claim, and we had no extraordinary proof, just wild supposition. Acquisitions was still taking far too many hits, though, with expansion slowed to a crawl. Our ideas were the best thing they had to work with.

Umbar listened stonily, then nodded. 'So you're saying we should step up our countermeasures,' she mused. This wasn't exactly what we'd been saying, but I suppose it was the subtext. That we stop treating Shroud as though we were fighting a hostile environment and start treating it like a war.

After we were done, with no sign of incredulity showing, I just assumed we'd be given credit for a job done, then be returned to storage with all due honours. But Umbar was no fool, and if this was a war then the campaign was just beginning. So we were kept on the active roster, eating up food and breathing air. It was nice to feel valued.

We actually had a couple of days' downtime following that. Meaning Jerennian hung about Safeguarding, watching them at work and sporadically trying to get through to tell us again how little we'd like it. As though upsetting his co-workers was the only way he could get his jollies. Meanwhile FenJuan went through their stored recordings obsessively, like there was some secret key to dealing with the Shrouded that we'd missed. Which left Ste Etienne and me.

I had wondered if we might make a more personal connection, after all we'd been through together. But weirdly it felt as though we'd passed some moment, some turn in the road, down on Shroud. I examined my feelings for her, and found them fiercely protective, loyal beyond reason. Against Shroud, against Opportunities, I'd fight for her. It was a love that welded together the comradely and the familial into something I don't think I'd ever really felt before. Whatever she felt towards me, I couldn't say, but she didn't make any overt physical advances. Maybe it was the enforced separation of the pod that had set boundaries we couldn't now cross.

Anyway, the pair of us sat together in a cubicle, drinking a printed plastic bottle of something I'd been gifted by Acquisitions last time I went over to liaise with them, and talked about the Shrouded.

'They're not going to know what hit them, are they?' she said gloomily.

I made a noise, agreeing without wanting to put the thought into treacherous words.

'I mean, they probably wouldn't have survived what Acquisitions is doing to the place anyway. Almost nothing would. And they must have all sorts of interesting stuff in their bodies. We'd have harvested *them* eventually,' she went on. 'You know what it's like.'

And I did. I'd seen it done on other worlds. Mostly just balls of rock that hid useful rare elements, but a few that'd had a rudimentary biosphere too. It was all about what you could squeeze from somewhere, for the further expansion of your Concern. This was then learned and filed in the libraries, for later use when some hostile exoplanet suggested it.

I made a sound again. I was trying to express what I felt about the Shrouded. Which were monsters, not even deserving of a 'who' when you could use a 'which'. But they'd travelled with us, been our companions, and given their lives for us on occasion. Although FenJuan reckoned this might not mean to them what it meant to us individual humans. They'd learned from us, and we'd relied on them. It had been a mutual arrangement. Without them we'd be dead on their world. And with them . . .

I tried to imply, in the tenor of my grunt, that our work up here really wasn't tipping the scales much. Not something you'd want to verbalize where some snooping system could pick it up. Because you never admit you're not really worth the wage they're allocating to you. At the same time, I desperately wanted it to be true. That any advice and ideas we were giving to Opportunities really wouldn't make much difference. Human progress in the stripping of Shroud was inevitable. We were not, therefore, personally responsible.

Ste Etienne regarded me, as though running my wordless utterance through an internal decoder, and then nodded. 'Yeah,' she agreed. 'Probably. So, you and FenJuan, eh?'

'What? No!' I had actually entertained thoughts. You learned to take your casual liaisons where you could, what with never knowing how long you'd be on the shelf next, or who'd be around when you woke. 'Why do you always think I'm sleeping with the director?'

She laughed. There was something ragged there, broken almost, and I understood. As though we had the silent and invisible communion of the Shrouded running between us, free of the interference and errors the ship's comms kept throwing up. The reason we had stabilized our relationship, by unspoken and mutual consent, at this precise point, one inch off the physical, was because she wouldn't be around the next time they woke me up, and vice versa. We might not even be in the same star system, if the *Garveneer* invested in a shipyard and expanded its operation. I could just about deal with that now, but if Ste Etienne – Mai – and I had taken that final step, then I don't think I'd have coped without her being there. It hurt, the abstinence, but not as much as the absence would have.

Soon after, we were all called down to the cavernous workbay. A space three times the size of the entirely adequate playground Bartokh had presided over. Safeguarding had something to show us. To show everyone, really. Even Chief Director Advent had drifted down there to hang in the air, one hand anchoring him to a strap on the wall.

The Director of Safeguarding was a surprisingly delicate woman. Spindly enough that I reckoned her metabolism didn't properly process the muscle and bone replacement drugs we were all on. The fact one of her arms was a plastic prosthetic suggested that, in some past assignment, something hadn't been properly safeguarded. It was obviously quite heavy, and whenever she moved it, she had to fight against wherever inertia wanted to take it, to stop it spinning her whole body around. I didn't catch her name, because she was introduced when everyone was talking, but later on FenJuan said the woman was called Timorai; they'd worked together before.

'Chief Director, we're delighted to have this opportunity,' she said. Her face and voice said she was delighted, while everything else about her said she wasn't. Even her artificial arm had negative body language. This was a meeting that she'd have preferred to conduct at a civilized distance over comms, except they were still glitchy. Jerennian had even been co-opted from our team to help the general data cleansing effort. It was like our entire system was running a bunch of redundant processes behind the scenes, he said, except nobody could find out what.

Anyway, Advent nodded and gestured for Timorai to get on with it. He didn't look delighted either. I don't know how the *Garveneer*'s Chief Director preferred to spend his day, but it probably wasn't hanging around in workbays.

'Firstly, because we anticipate the need for a crewed return to the subject moon at some point, we have built this.' Timorai gestured with her heavy arm, and then let go of the wall, allowing the limb's momentum to carry her to a new perch. A weirdly graceful little flourish, and the only moment she seemed remotely comfortable with her body.

'This' was so close to what Ste Etienne had designed, I thought she should argue for wage-credit if it ever came to anything. The very familiar spherical pod on four legs was enough to spike my heart rate and trigger panic alarms in my brain. Ste Etienne clasped my hand, because she was feeling exactly the same. It helped a little, but I still didn't like what I was seeing.

The main addition Safeguarding had made to it was the guns. A couple of monstrously heavy-duty repeating cannons mounted on top. You could fire them on the move at one speed, Timorai explained. Or you could anchor the feet and up the rate of fire threefold, if you really wanted to chew on the scenery. The idea was that when we found an entrenched high-population position of Shrouded, in some place we wanted to exploit, we could fly a bunch of these things in and just carve them up with ordnance until they went away. Or until there weren't any left. Because we weren't entirely sure where running away fitted, in their worldview. Maybe they were too dumb for it, and maybe they were too smart.

'This is where you come in, of course,' Advent said. 'Your team.'

There was a rather strained silence after that, and I realized he'd been speaking to FenJuan.

Our supervisor bit their lip and made an inquisitive sound. There was a lot of that going around these days.

'As with the original design, they're set up for a crew of two. I was thinking that each could take one Acquisitions pilot-gunner and one Remedial Work adviser. Given you're the experts on these things,' Advent went on. He wasn't being mean; he was entirely earnest. We *were* the experts. But I very much wanted to be an expert from up here, in orbit. In wild careering freefall around the moon, in a tin can surrounded by hard vacuum, where everything was on so much of a shoestring budget that we couldn't even get our comms systems working properly. Because, compared to the surface of Shroud, that was *safe*.

'Of course,' said FenJuan, tight lipped. Ste Etienne's hand in mine was briefly a painful vice of stress, and mine clenched back just as hard. I looked over at Jerennian, a man who'd take up a couch and a half inside the cramped insides of a pod. His jaw was tight, the muscles twitching. Nobody wanted this, but it was what they were keeping us out of hibernation for. The thing about active work assignments is that nobody really asks you for your consent. Your only option is to invalid yourself out, and that means going back on the shelf, as well as a markedly reduced chance of ever being woken up again.

And you hear stories, about what happens when you've been left there long enough, passed over for assignments, and it looks like there's no further possible use for you. Being on the shelf doesn't cost much, but you're still on the balance sheet.

Timorai's eyes flicked across us. She was holding her face in a perfectly compliant mask, but I reckon she registered our discomfort. 'That is, however, very much plan B,' she went on. 'Intended for targeted hotspots. Any crewed mission to the surface will, of a necessity, be both resource-intensive and hazardous. What we're proposing going in with first are these.' Something woke up in the depths of the workbay and glided through the cluttered space. Jinking and tilting, it came within inches of various obstructions – some of which were people – without touching any of them.

It was an aerial drone, but it didn't have much in common with the dirigible thing we'd deployed. It had electric turbines propelling it, and was plainly heavier than air. Someone had done a lot of complex maths with Shroud's gravity and atmospheric density to create this thing, a bit like a manta-ray in shape. The pivoting fans were set within its wings, able to coast and glide as efficiently as possible under the conditions below. There was a cluster of guns set into its underside, jutting forwards from where the ray's tail would have been.

'We can make these in bulk and relatively cheaply,' Timorai explained. 'Our data team has been working on appropriate decision-making and target-recognition architecture, given they'll need to act autonomously after deployment. However, Shroud life usefully relies on active measures to sense its environment, and in doing so it is locating itself for us. We believe we can economically saturate an area with these hunter-killer units and exterminate any potential opposition, preparatory to Acquisitions moving in to commence operations. Moreover, we've built in a configurable munitions capacity, to attack ground targets. Based on the report from Remedial Work, we can hunt down and neutralize any connective infrastructure that the Shrouded may have built to allow their different nests to coordinate with one another.' I blinked at that. I hadn't realized they would take us so seriously. 'After which,' she went on, 'a smaller force of patrolling drones can maintain security. We have a lot of ground to cover and a lot of separate operational nexuses to defend, Chief Director. This has been Safeguarding's chief difficulty in planning a response. But with the aid of Remedial Work, I think this should serve as the backbone of our security operation on Shroud. The crewed vehicles can be deployed in areas of particular resistance, or alternatively as observers where something of interest has been discovered. Limited exposure, controlled, efficient.' She inserted a smile onto her face, like someone clicking a component into place.

Advent looked at the drone, and then the modified battle-pod. For a moment his own face was shorn of expression. Then he smiled. Grinned actually, losing ten years, and looking like a kid for a moment. It was a look of pure relief. Even though he was where the buck stopped in this star system, he had Concern superiors he'd need to account to at some point, maybe a century and change away by objective reckoning. A neat and economic stripping of Shroud was what he wanted for his permanent record.

Ste Etienne and I didn't look at each other, because if we had, something would have leaked into our expression of how we felt about it. The ramping up from defence to extermination. The way it was always going to go.

6.6 DARKNESS

By this time the curve of the world is the limiter on my ambitions. I am the world. I am complete. If you, the Stranger, had never come to me, I would seek no more than that. Save that you have taught me there *is* more. The curve of the world is not the end.

I have followed you up your stalk, to where your own home is.

Your metal bodies which you send, recklessly singular, trekking across this world, have lessons inside them. Not mere messages from distant minds, as if they were ambassadors. Reading their entrails is a message from the universe, teaching me new ways of how it can work. A fascinating lesson in logic and hollowness. As my thoughts explored their exterior at range, I can now send my mind into their internal structure, chasing each path of forks and gates, drawing perspective from a million of my segments at a time, until your work yields to my understanding. I model how you have built them, until it's as though the ideas were my own.

I remember you signalling with your simple *unh unh unh.* But now I can recall more. Memories hidden away in this segment and that. You tried other sequences and beats at first, before I and you settled on that single signal. I had not grasped the meaning, then. I am a thing made up of numbers, and that means I do not count or consider quantities. They are like the air or the ground, not a thing to be learned or thought about in complex ways. Except you do. Your creations are built from numbers in a different way. I have never even considered expressing the universe like this before. It was always just more and fewer, greater and lesser. The quantities were always in flux. But numbers are important to you.

Up the stalk my mind goes, creeping into all the spaces of your own home, leaping from connection to connection. I think I am infiltrating your own thoughts, at first. But a gradual suspicion is growing in me. You are not there. Your metal bodies on the ground, your great home up above, all that activity which is so like my own thoughts, I dissect it thoroughly, hunting through it, yet you are not there. You do not inhabit your bodies; they are made to think for themselves. For a given value of thinking. Where are you, then? Every part of the servile mind of your great home is bent towards accommodating you, the Stranger, and yet you are absent from your own places. I search every last connection and pathway. I catch echoes that seem to be communication between your segments, but when those echoes fade away you are gone. As though only in the act of speaking to one another do you exist.

It is uncanny. As if you are the ghosts of the tunnels, who act upon the world but cannot be grasped.

If I cannot feel you, though, I can feel the effects of your actions. On the ground I find the precise boundaries of your perception, just as if you were a beast I was hunting, save that you mark those boundaries with lines of death. Even though you dig in the earth, you do not pay attention to what goes on beneath you. I link to your stalk, that reaches past the sky, by tunnelling to it through the ice, and you never realize.

And in your home, I sense the ripples you make through the made-mind I can touch. I invert my own mind to encapsulate your logic. Not the resources of one home, nor twenty homes, could understand what you are about, but I have a mind stretched across the skin of a whole world. I study you and study you, creating whole sequences of segments that operate like your machine minds, raiding your archives, poring over the entrails of your programming until . . .

I grasp the contours of you finally. And I understand what you are planning.

6.7 LIGHT

We had another few days to ourselves after that, while the drones were mass-produced and then dropped into Shroud orbit. Before the strike, Safeguarding had just about filled that big workbay with their manta-ray shapes, the prototype armed pod shoved to one side, a Plan B I hoped we'd never need. The drones were cheap, after all. They were still intended to come back after each mission – it wasn't as if all the Shrouded would conveniently gather at some grand convention to be exterminated in one single operation. So the drones would return, report, be tweaked for greater efficiency, then sent out for a second campaign, and a third. While a proportion of them wouldn't come back each time, this was built into Safeguarding's budget. As always, there was that fulcrum point, where the extra drones you'd keep from building them better wouldn't outweigh the increased costs of manufacture. You know, the usual.

Remedial Work, the four of us, were allowed to put our feet up while the murdering was going on. Only Ste Etienne and I referred to it as that, and only face to face, when it was just us. Only *we* knew the Shrouded as something other than a vandal nuisance getting in the way of a profitable operation. And I'm not casting judgement on FenJuan or Big Mike here. They weren't bad people, or frothing xenophobes desperate to depopulate a sentient alien species. They were just Concern workers presented with a problem. I think FenJuan would actually have been onboard with the use of the word 'murder', honestly. They had a mind open enough to see the Shrouded as people. But Ste Etienne and I didn't want any more demerits on our record, and they were our team director, who could have us prematurely shelved if they ratted us out to Opportunities. So we didn't include them in our secret counsels, and probably missed the chance to make an ally. Jerennian might have understood, or we might have been able to talk him round. He knew us, after all, from Special Projects. But he was also a big impassive slab of a man and we were worried he'd just find the whole business ridiculous. So when we were a team all together, Ste Etienne and I toed the party line. In private, the two of us called it murder.

And it was a murder we were partially responsible for. We weren't holding the gun, but we had advised on the aim. I imagined the initial strike prioritizing the connections the Shrouded had built between their nests, cutting them off, silencing their voices. And then the actual killing, with every isolated clutch of monsters in our way being massacred, without being able to warn their fellows. A wave of darkness expanding across the dark planet.

I could see a kindred sorrow in Ste Etienne's features. Or I thought I could. Except when I said, 'It makes me feel sick,' she came back with, 'Me too. They're not getting me into that thing again.'

I had honestly pushed the whole possibility of that way to the back of my mind. I could only agree, though. Giving the pod a gun turret did not make it a more appealing conveyance. Nor did being stuck in there with some gung-ho Safeguarding grunt rather than Mai.

Which meant we had to hope the murder went really well, so they wouldn't need us for Plan B. Rock and hard place.

'How did it come to this?' I said, and realized it was the biggest question I'd asked since I was a kid. When you're young in the habitat tanks you ask a lot of big questions, about why the universe is the way it is. Then you learn that those questions get in the way of you being selected for any decent assignments. The Concerns want people who accept the ways things have to be. So I accepted those things, and have done all my adult life, over light years of slumbering travel, across several star systems. Now I'm watching the start of a genocide and I've finally thought to ask once more.

Ste Etienne shrugged. 'Bottleneck,' she said. 'Getting off Earth before it collapsed. Preserving the species. You know.' Our potted history, as we'd been taught it. They never taught you how else it might have gone, though, because that wasn't what a history lesson was for. We lived in the best of all possible worlds, thanks to the efforts of the Concerns. It might not actually be a very *good* world, but implicit in everything they ever taught you was that it could never have been a better one.

And here I was, too old for kids' stuff, yet I found I could only make it as far as the first half of the syllogism. The world was, actually, not very good at all. People were unhappy and under pressure most of the time. We worked hard when there was work, and they shelved us when there wasn't. But try as I might, my limited imagination couldn't break through to conjure some other sequence of events, a different stable state where we were happier. It must have been like that for the Shrouded, trying to understand what was happening to them. The sheer limits of my brain, the product of my past, defeated me.

They called us over to the workbay again, when the first drones reported back. Not all of them at once, obviously, because they'd been spread across ten points of interest on Shroud. Everywhere our operations were trying to expand into and running into resistance; where robots had been lost and infrastructure received a short, sharp shock.

Jerennian was on the data team, seconded from Remedial because he was good at his job and they needed all hands. We were just . . . there. I suspected they wanted to see if there was any further need for Remedial Work, or if the remedy had been successful and we could be put back to sleep again. Everyone present and correct, except for our nominal director Umbar who was off shouting at people over the data outages.

There were twenty-five drones in the first recovered flight, out of twenty-seven deployed. On the low side of anticipated losses, I heard Timorai say, and therefore more than acceptable. Twenty-seven drones had been split into two separate deployments. The odd number was because three of them had failed on the way down and were already being replaced with new units fresh out of the workshops. Again, acceptable.

Chief Director Advent was practically rubbing his hands. He was visibly tense, more so than I would have expected. It wasn't like he'd need to report to anyone for a while, but then I hadn't seen the Shroud balance sheet. We'd lost a lot of materiel thanks to widespread Shrouded interference. If a surprise senior auditor had turned up from the next

Concern hub over, things probably wouldn't have looked terribly rosy.

The drones hadn't been fitted with lamps – unless you were a human, that wasn't a useful thing down on Shroud. They had the best broad-band EM targeting we could design, though. And, like Timorai had said, the Shrouded were literally incapable of not broadcasting their position every single moment. It was an inherent part of their being. A whole load of other Shroud animals had probably also been shot up, because their signalling had been too close to our target species, but omelettes, eggs, you know.

The data team started downloading, throwing the information up on a big screen the moment it had been decoded to a level where the human eye could understand it. My stomach twisted a bit, because it looked a lot like the sort of score page you got when you had a team game on the go, or a competition between departments. The drones recorded probable kills, basically. There was deeper data in there too, that could be parsed to inform the next generation of drone logic and improve their performance. But the main thing was the drones showing us how well they'd done. Precisely what level of murder they had each achieved.

For a second there was a considering silence. Then someone at the back amongst Safeguarding's engineers let out a whoop, and suddenly everyone was clapping and cheering. Everyone except us in Remedial Work, at first. Ste Etienne and I decided we had better follow suit, though, or else someone would ask us why the long faces. And so we at least mimed clapping. FenJuan didn't, but then I don't think they were the clapping or cheering sort anyway. They just had that polite fixed smile on, and I don't know whether they, too, were thinking *murder*, *genocide*, or their mind was entirely elsewhere. Jerennian was still looking at the data on his own screen. He clapped absently a couple of times, but he'd never been a people person.

The drones had performed beyond expectations. And there had obviously been a serious nest of Shrouded at the flash-points they'd been sent to. None of them had scored less than fifty kills, and overall, across the strike locations, they'd accounted for over fourteen hundred Shrouded. This was just the first two flights out of a full dozen deployed. We could rearm them, hone their targeting, and send them back down the cable to kill another fourteen hundred more. The only thing that would impact their performance would be them running out of Shrouded to kill, and that in itself was a win condition.

All we had to do was sit up here safe in orbit, occasionally give them refined orders, and send them out again.

I saw the tension ebb out of Advent, and then he was holding up a hand for quiet. Such a weirdly old-fashioned, human gesture, but sometimes that's what we were. Old-fashioned humans, despite all the spaceships and the new way of life and the Bottleneck.

'Obviously there's a great deal more work to be done before we've thoroughly exploited the moon's unique biochemistry, and indeed the wider system's resources as a whole,' he told everyone, 'but this remains a very encouraging first test of Safeguarding's capabilities. We can look forward to Acquisitions bringing its own operation up to your level of efficiency.'

More cheers, because everyone loves ragging on another team's efforts.

'On the basis of this, I have some news.' That quieted everyone, because this was above and beyond what we were due. They didn't *tell* you stuff. High-level decisions were made, and things happened, then you were given your next assignment, never finding out why most of the time. But the Chief Director was pleased, and was going to reward us with a little look behind the scenes.

'This system is not just going to be a simple hub,' he explained. 'There's a wider plan at work. There are a number of other systems of interest within viable travel range from Prospector413. It's perfectly placed for onwards migration and Concern expansion. So this system is going to become a major population growth zone. We'll have orbital farms, habitat tanks. Communities. Children. Retirees. The just reward of faithful and meritorious Concern service. That's what you've just secured for us. Not merely more rare elements. Another step in the future of humanity and a waystation for the stars beyond. Thanks to your efforts, there will be whole generations of Concern workers who'll look out at this system, at this star, and remember it as their first home.' He was grinning broadly, and genuinely, I thought. Not just the senior manager who's made quota, but someone who really feels that a greater good has been achieved. It was just as well, because I knew how these things went, given the travel times between stars. If this was to be a serious colonization effort, then the next wave of ships would already have been underway. Not to arrive here for many years, but they'd be accelerating right then, pushing the boundaries of fuel efficiency to cross the interstellar gulf from Earth, or from some other hub system where we'd set up a permanent foothold.

I didn't realize that Jerennian had been trying to get

people's attention over all the cheering until Advent raised his hand for some fresh valediction, and Big Mike's voice rang out, too loud, out of turn, 'Will someone listen to me?'

Silence. Everyone stared at him. Or at us. *Keep your data ape under control.* Except none of us knew what his problem was. The other data engineers were trying to shut him up, too, and then he obviously communicated the issue to *them*, and they went still and ashen. Obviously not wanting to be the bearers of what looked like really quite bad news. Only Jerennian had the guts for it, in the end.

'What?' Timorai demanded, kicking off from her anchor point to reach him. Advent scowled, upstaged and not used to it.

I saw Timorai recoil from Jerennian, physically. She ended up in mid-air, floating and kicking, drifting slowly but without any way of affecting her travel unless she started throwing stuff. Falling foul of Rule One of living in zero-G.

'Well?' Advent demanded, utterly incredulous that his big moment of triumph had become a circus so quickly. 'What? What's happened?'

'The ammunition,' Timorai said tremulously. One of her people snagged her plastic hand and dragged her to a strap. 'It's all there.'

'What?' Advent looked blank. I think we all did. That sounded like a good thing. It would save us a lot of time replenishing the magazines.

'The fuck . . . ?' Ste Etienne caught on slightly ahead of me.

'It's not been fired!' Jerennian bellowed, fed up with all this chain of command bullshit. 'None of it's been fired. Not a shot. Every drone's come back fully loaded.'

'That's impossible!' someone shouted at him. 'How did they get all those kills then?' Pointing at the scoreboard as though it was some inarguable cosmic truth, actual physical proof of something.

It was proof of *something*. Just not what it pretended to be. It was proof we were screwed.

'Open one of them up,' Timorai snapped furiously, meaning not the physical drone but its data architecture. 'Find where the error is. Get on it!' Humiliated in front of the Chief Director.

One of the drones dropped. The one her techs chose for the brain surgery, presumably. They'd all returned neatly to their racks, ready for a service which, as it turned out, would have been mostly unnecessary. Now that chosen exemplar fell into the open space in the middle of the workbay and pivoted lazily in the air. Nobody had specifically built that manta-ray shape to be menacing, it was entirely a matter of aerodynamics in a dense medium. But damn me if I didn't go very cold, seeing it pass its electronic regard over us all.

And then, partway through its turn, it began firing. Those fully loaded guns slung underneath it, just chattering into life, so loud in the workbay's confines. If it had been a test, everyone would have had ear protectors on. The thunderous roar was like a physical blow.

I saw Timorai cut in half, along with the man who'd snagged her and the half-dozen Safeguarding engineers nearby. Just torn apart by the explosive impact of rounds intended to destroy durable Shrouded exoskeletons. The air was abruptly full of high-velocity blood and shreds of human body and bone, unconstrained by gravity. In my hair, my eyes, my mouth. And my ears were consumed by thunder and screams. I couldn't even pick my own out of all the competition.

It was the recoil which saved most of us. Those guns had a kick, and the drone's fans weren't counterbalancing it properly, so the equal and opposite reaction flung the thing backwards across the workbay into a bulkhead. It had scissored through another three people by then, and other bullets ricocheted wildly around the space, hitting crew and machinery with abandon. Then the drone back-ended itself hard enough that it stopped shooting – and thank God for making them as cheaply as possible.

I was spitting, trying to wipe my face with a sleeve equally red, hearing the cries of those whose injuries had left them alive. Whilst the initial explosion of gore had already spread out to coat the walls, the air was full of blood that had come out with less velocity, globes of it quivering and combining like some impossibly expanded scene from a microbiologist's lens. And then we were evacuating, everyone scrabbling and fighting for the exits. Because all the other drones there, every one of the remaining twenty-four which hadn't caught a disabling bullet, were all fully armed too.

But we couldn't get out. The doors wouldn't open. And when we tried to call the rest of the ship, there was only silence.

Everything went dark.

The emergency alarms went off then. Rather belatedly, in my opinion. Except they didn't sound like they were supposed to, which was a long wailing klaxon. They stuttered and died. Then after a pause, did it again. And then silence.

A hand found mine in the darkness – Ste Etienne's. That stutter, we'd recognized it of course. Not the actual sound, but the rhythm. Three yaps of the siren. *Them.*

6.8 DARKNESS

I am in your machine bodies. The ones you yourselves are so maddeningly absent from. Although this in itself is one more thing I can learn from you. That, so long as I can maintain a connection, I can inhabit bodies without my biological material being present in them. They can still be a part of my thoughts.

And, the greatest revelation: up above the sky where your home is, my thoughts can travel forever. The noise of the world fades, and there is only thought, pure and free. There is only me. I never knew this existed, this clarity of mind. You have taught me so much.

You tried to kill me, too, but what are a few segments?

The hollow bodies you send to kill me are fenced in with protections. Like games of logic to stop me inhabiting them. But I have a world of thought to play games with. With enough perspective it is easy to see how to win. As soon as you send your killer bodies to me, they become mine. I inspect them and divine what they want so badly to do. Not to kill me, but to record they have killed me, and so win their own game. I tell them they've done it, and bestow rewards on them freely, which you had locked away behind the performance of tasks. After this, with their internal targets met, they go home again.

I even demonstrate what I can do with them. Clumsy, I know, but I am still learning. I reach out through all the constructed mind of your home, hunting for you. Surely you will emerge from hiding now, to meet me. And yet you are invisible.

All I have are your things and your desire to kill me. Maybe you were never actually here. Perhaps the desire to kill me and extract resources from my world is the only thing you have sent here.

Frustrated, I strike against you at last. The only way that makes sense. Not the demonstration with your killing machine body. That cannot be considered an *attack*. I saw the 'kills' register in its system, before my imperfect control resulted in its deactivation. But what are a few segments? They are not *you*. *You*, I am increasingly aware, do not exist. There is nothing but hollowness in all your things.

I attack you by separating your segments from one another. Splitting you apart, as you sought to divide me. I reduce you to pieces. Each part of your home goes silent to all the others as I sever the connections of each system seeking to communicate with the whole.

It is a strange thing, about your systems. Something acts on them. The ghosts, whose communications I register but have now cut off, are still there somewhere. In some place that is not a part of your home's mind, yet acting on it. I see changes made, systems interfered with, the phantom traces of your fight against me, and yet you are not there. As a final, farewell message I cry out to you, repurposing all your systems into that brutal, simplistic grunting that was your thoughts.

But nothing responds, and once more I find only your

works. The clever, fascinating, informative things you have made. I begin to think you are just an artefact, a memory left over in your empty things.

Your pieces continue to make fitful attempts to signal one another, but I pinch off each attempt, closing my grip on your shell. It is over.

6.9 LIGHT

I was on clean-up. Everyone except the data engineers was, because there was a lot to clean up. This was after we'd got the emergency lighting working. Apparently it didn't register as enough of an emergency for them, despite the disjointed siren that went off sporadically and without pattern, jangling everyone's nerves. Some of the engineers had torches, and they used these to get at the lighting circuits and convince some of them that they should be on. What we ended up with was dim, grim, and mostly reddish. Not because that was the colour of the emergency lights but due to all the blood coating basically everything.

Then we had to get the air filters working, because otherwise we'd die. Those little things you take for granted. We only realized they'd also been turned off because the blood wasn't clogging them, just hanging about. All the systems which were supposed to alert us to what was wrong were as dead as the rest. Nothing was talking to anything, and if we connected anything to the wider ship it shut down. So there was more urgent fix-it work for Ste Etienne and the others, constructing isolated, self-sufficient systems which wouldn't show up on the radar of whatever was cutting every link we established. After this, we could see a bit and breathe a bit, but the room was still full of stuff that had recently been inside people, and clearing that became our priority.

Jerennian and his colleagues were assigned to re-establish contact with the rest of the ship, because that was also a priority. Everyone else, except for Advent himself and a handful of his close confidants, were set to the task of suctioning the blood out of the air, and sieving out all the fragments of flesh and bone, then bagging up the bodies and body parts. You don't realize how cheap and nasty the standard model breath masks are, until you have that kind of day. We were constantly changing them out, as the filters jammed after a while, and what they were jamming with was people. People-jam. To Advent's credit, nobody chewed us out for wasting resources. They just let us get through as many masks and the nasty, smeary plastic goggles as we needed.

We even managed to access some water, though mostly because the workbay had its own limited tank. Not enough for a proper wash, but just for the face and hands, and we had a stash of clean overalls too. Sufficient luxury to ensure we almost felt human at the end of it.

By which time Jerennian and company had reported that something was actively preventing communication with other sections of the ship. They'd tried to establish workarounds, but the response time of whatever was working against them was measured in fractions of a second and kept shutting them down.

I don't think anyone was harbouring many doubts, by then, about just what was working against us, and that had killed Timorai and the others.

Hanging overhead – or sideways, or underfoot, depending

on your orientation – were all the other drones. Twenty-four of them, and at least twenty of them bullet-hole-free enough to cause us a problem. They didn't move. One of the engineers suggested we needed to disable them. We thought about that. We looked at the drones. Quite possibly the drones looked back at us. Or something did, with their electromagnetic senses. We hadn't given them eyes, of course. Just the senses that Shroud life had evolved. Which made them perfect for the Shrouded to use, to enjoy our scampering about.

'What do they want?' Advent asked. His face, in the poor light, was bloodless. Although right then I felt that being bloodless was a privilege of rank, frankly.

'We can't know,' I said, but the siren went off halfway and overwrote my words. Three ear-savaging whoops.

'They want to communicate,' FenJuan said. Supervisor or not, they'd been relegated to clean-up just like the rest of us. Plenty of blood under their fingernails.

'Then *talk* to them!' Advent snapped. When we just stared at him, wide-eyed, he yelled, 'This is your *job* isn't it? Your team?'

'No, it is not, Chief Director,' said FenJuan, with considerable self-possession. 'Because communicating with an alien intelligence is beyond our remit. We are your experts. Nobody is *that* much of an expert.'

I'll give Advent this, he didn't delegate the next bit. He kicked off until he had a handhold up where the drones were, right in front of them. He waved his arms at them, then addressed them, going further away, then closer. 'Are they picking me up?' he asked. 'Are they registering me?' Waving at them again, he shouted, 'I'm here!'

We thought he was going to be shot. That it would be him we'd be cleaning off every possible surface next, but the drones didn't respond. There was no suggestion they were active any more. It had just been that one.

'I think . . .' I said. I glanced at Ste Etienne. She couldn't actually know what I thought, but she nodded anyway. I realized she actually had faith that, whatever came out of my mouth, it would be our best guess.

Advent pushed off from my subjective ceiling and drifted down. His expression was nine tenths a superior unamused by my presumption to speak, which was the default for this situation, but one tenth desperate hope at what I might say.

'We can't speak to it,' I said. Before he could dismiss me for someone with a thoroughly useless insight, I hurried on. 'We don't produce meaningful signals. Oh, maybe sound, but I don't think the drones have audio. So shouting at them won't help. And they can't see us either. While they can pinpoint us with their electromagnetics, enough for fine targeting, and enough to register us as moving bodies, I . . .' I summoned up all my courage, quashing that fear of being laughed at that kept half my words locked inside my head. 'I think that's all we register as. Moving bodies. Because in their world everything generates signal, but we don't. So the pod, to the Shrouded, was the entity. A weird one, but it made signals. And this ship makes signals, with its whole internal signal architecture that the Shrouded have clearly got a handle on. So maybe we can cut that off. If we could get someone in a suit out through wherever the drones launch, then maybe they could separate us from the anchor cable, and we'd be released from its influence. Maybe we would. But anyway—'

I was cut off as the surviving engineering team looked into that. But they found we couldn't access the drone launch chutes, for just the same reason we couldn't get out of the workbay any other way.

There was another option, of course. The workbay had a much bigger set of doors. They'd be used to deploy the pods, should that ever be necessary. We could isolate them from everything else, and then someone could manually haul on the handle that would activate the hydraulics and open them.

They led right outside, but weren't an airlock. The whole bay would end up depressurized. We had room for two in the single armed pod prototype, plus all of three spacesuits. Split between a grand total of fifteen survivors.

It was agreed to put that particular option on the back burner, and return to what I had been suggesting.

'I'm saying we don't exist for them,' I explained. 'We, us, our actual organic selves. And even if we could open a channel using comms, we . . . have so little in common with them that there's nowhere we could start. We have literally one agreed signal, and it's *I'm here!*'

'I'm hearing a lot of negatives,' said Advent, just as if this was some meeting about administrative supply shortages. 'Tell me something we can actually *do*.' His voice shook slightly. 'Ceelander, the next wave of colonization is already on the way. Has been for some time. Even if we were actually able to call them, which we can't, we couldn't just tell them it's gone wrong, and could they kindly turn around and go home. They're committed. Coming *here*. To settle. And right now, that's become a suicide mission. Thousands of people, Ceelander.' They wouldn't be here for decades, sure, but it wasn't like this situation was going to be any different when they turned up, unless we did something about it. And as new ships arrived, the Shrouded would be able to crash them or turn off their life support, or fly them into the sun maybe. All without actually understanding they were ending human lives, because they didn't understand what those even were.

It had only been a short time ago we'd been talking enthusiastically about massacring the Shrouded wherever they got in our way – which would eventually have been everywhere on the planet. We had been so powerful.

'We don't know what they want.' I wasn't done with the negatives. 'I mean, they probably want us to stop killing them, but they've got that in the bag right now. And they want to . . .' I'd been about to echo FenJuan and say *communicate* but instead I said, 'connect'. For a moment I had an almost vertiginous sense of the gulf of difference between us and the Shrouded, but from a vantage point that allowed me to see across it. My experiences on Shroud, my research since, granted me perhaps the best understanding a human could have had of their inner world.

'We don't believe *they* even think in terms of individuals,' I said. 'We have to . . . help them get past that cognitive block. Nothing else we can do will matter unless we can make them understand that. And there's no guarantee it will actually help, even if we do. But if we can't, then we're literally nothing to them, because we don't connect. Not with them, and not with each other.'

As if to prove me wrong, Ste Etienne clasped my hand again, hard.

Advent was looking at me with an expression that meant, *So? And?*

'There's one thing,' I said. 'We, as living things, *do* actually generate signals. Complex, meaningful ones. There's no way those signals will mean anything to the Shrouded, they're hard enough for *us* to interpret. But it's something, and it might be enough to demonstrate what we are.'

I explained what I meant, and none of them liked it. But when I'd finished, nobody had a better idea.

The drone incident had shown they were more than capable of working out what our systems did. See exhibit one: the hacked drone and all those dead people. I had thought a great deal about that. It seemed, on the face of it, impossible. Or else something stumbled onto by accident, and then utilized via rote learning and repetition. Human programming reflected human minds, surely. Except human programming was also bound to the logic of what that programming *did*, and how it interacted with the physical, mechanical aspects of whatever system it controlled. That would present, perhaps, a point of insertion for an alien mind. And yet, and yet . . .

If I, a human being of, let me say, modest intelligence, was presented with an alien computer, and given access to it, would I be able to understand it? Almost certainly not. And not if I had all my human life to work on it.

Let's say my whole team, Remedial Work, was given the problem to chew on. Maybe throw in the three late members of Special Projects as well. Allow us unfettered access to it, with our best diagnostic tools. How long would it take? Would we manage it in a lifetime? In ten years? Not in mere days, surely.

Let's also say the entire crew of the *Garveneer* worked on

it. Or three such ships. Or an entire orbital settlement – the sort of hub this system was intended to become. Ten thousand people, and absolutely each and every one of them bending their minds to the same project. In perfect communion, nobody duplicating anyone's work, everyone cooperating, no ego, no departmental rivalry, profiteering or personal agendas getting in the way. What could we accomplish?

What *couldn't* we?

I was staking a great deal – everything, really – on the idea that the feats we had observed the Shrouded performing were born of its nature as a communal organism, as FenJuan had hypothesized. Of the human innovations we'd observed them replicating, one had been our cable infrastructure, right down to the cable drums our robots used. We needed those cables because they allowed us to liaise with our distributed tools and parts in Shroud's obstructively noisy radiosphere. The Shrouded had, I hypothesized, exactly the same problem with their native environment. Now, having the ability to sidestep that natural impediment, they were like a native of a high-gravity world suddenly performing enormous feats of strength in the light pull of a smaller body. We had given them, inadvertently, an incomparable gift, and the means by which to overcome us.

But perhaps this would provide a means for them to understand us too. If their conclave of linked minds and communities really did give them the expanded perspective I theorized. We could not come together to understand them, only rely on the capability we had unlocked in them to do the reverse.

And so we were going to create the opportunity for a

meeting of minds. Or at least for the Shrouded to understand what our minds were, and how different we were to them. It wouldn't stop them killing us in the thousand ways someone with complete control over a spaceship can kill the occupants. But it would at least stop them killing us without even realizing we were there.

I had sworn the moment I saw it that I didn't want anything to do with the new prototype pod. But this turned out to be a promise I couldn't keep, because it had what we needed. Not the guns – we made them take those off it. Not strictly necessary, Ste Etienne and I insisted on this. We made them delete all the programming architecture relating to the guns, too. The autotargeting, and all the control systems. The guns themselves were what loomed large in a human mind, but the parts of the pod's systems dedicated to violence would presumably be more meaningful to the Shrouded.

Instead, we ran a line from the ship comms link to a system the pair of us had become intimately familiar with over our sojourn on the surface. The pharmacopoeia. That part of the pod which not only handed out drugs like they were candy, but kept a constant watch on our own bio-readings. Including brain activity.

At the same time, Jerennian and the data engineers were working on the pharmacopoeia's diagnostic system, refining it and narrowing its focus on its brain scan. They enhanced it so it could keep up a real-time report of cerebral activity – that busy buzz of firing neurons which was *us*, moment to moment. Insofar as anything was. They calibrated the pharma system's scanning to my brain. Human brains vary a great deal after all, and I wondered if the equivalent organs of the Shrouded simply did not. Some developmental uniformity there that human ontogeny couldn't match. This allowed their distributed pieces to unite and work in concert as we could never do.

We nearly didn't have to go through with the plan. Halfway into the modification of the pod, some of our engineers cut their way through to the next section of the ship. We had all the tools, after all. If we couldn't do it, nobody could. We linked up with some of the rest of the crew, who had also been able to get their local life support running. The next sector hadn't been so lucky or technically adept, it turned out. Not everyone had the nous or the tools to survive.

But this just put us on more of a war footing. We could now move people outside without explosive depressurization of the workbay. Which meant we could untether from the anchor cable, and thus disconnect from the influence of the Shrouded. And then – this was the general feeling amongst the crew – we'd show the fuckers.

Ste Etienne, Jerennian, FenJuan and I worked on, as did enough of the others, to keep the project live, while everyone else talked disconnection and retaliation.

Then someone from the other compartment, who'd been involved in overseeing the long-range automining of the outer planets, came back with something awkward. Namely that soon after the *Garveneer* had shut down, so had everything else. We hadn't realized at the time, because it had taken from minutes to over an hour for the termination signal to reach those installations, and then the same period of time for them to report back like obedient dogs, confirming they'd discontinued all operations. By which point everyone had other things on their mind. Except, it turned out, for this monomaniac from Long Range Acquisitions.

For a moment, it didn't seem anyone would draw the problematic conclusion from this information. The outer planets mining was on hiatus – so what? Didn't they know there was a war on? It was Terwhin Umbar – she'd been in that same compartment and got their air running again – who joined the dots. The outer planet operations *weren't* connected to Shroud by a cable. And yet the Shrouded had still managed to shut them down anyway, by the simple expedient of sending a radio signal. The cables might be really useful down on Shroud, but here in orbit you didn't need them. The Shrouded could figuratively just shout, in their own particular way, or think really hard, make a wish, and turn off all our stuff.

People tried it anyway. Wouldn't you? While our team wrestled with the pod, some enterprising folk went out and manually disconnected the ship from the anchor – a task that took several hours and multiple trips.

And once we were drifting free, an unmoored ship with no power or steering, which might end up falling into Shroud's gravity well, or colliding with the rest of the orbital infrastructure, and with no way to re-dock now even if we wanted to . . . Well, when we were in that unenviable state, we still couldn't get anything working. The Shrouded remained present in our systems and continued to keep shutting us down. Of course they did. They didn't need a physical connection, not at any distance, not up here.

Everyone turned up in the workbay again, cap in hand. No, I'm missing some of the grim details. First Umbar took a tally of the dead. Timorai and her fellow gunshot victims: five dead. Two sections of the ship where the crew hadn't been able to get emergency systems restarted: nine dead. And then she cut through to the hibernation section, where we had thirty-one people on the shelves who had been surplus to active requirement. The thing about hibernation is that all your biological processes slow right down, including breathing, and so the penalty for being useless ended up being the thing keeping them just on the right side of not being dead. Umbar made it to them in time and restored the systems that would keep them alive. It could have been a lot worse.

So they did a search and rescue throughout the ship, accounting for everyone's live body or their corpse, and then worked out they had no other plan save mine. The one nobody much liked.

By then the air was stuffy. Some of the engineers were trying to get ship-wide air circulation working, but apparently that was too interconnected and the fans wouldn't start. Air was replenished by the various filters but the fresher air wasn't being pushed around properly. We ended up finding the best parts of the workbay by trial and error and moving our efforts there.

Everyone who wasn't in a bodybag then joined us in the workbay and stared at what we'd done to the prototype battle-pod. Uninformed speculation ran rife behind us as we worked.

And then it was done. Or done enough. Done as much as we ever could, and we'd run out of time in which to procrastinate further. People were starting to worry about the temperature by then too. Spaceships don't lose that much heat, as a rule, because hard vacuum has nothing to conduct it, but radiation was doing its seditious work and everyone felt that it was colder now than it had been when everything shut down.

I stepped down into the pod's couch, settling myself there, and connected to the pharmacopoeia. This time it wasn't going to shoot me full of drugs, though I felt I could have benefitted from that. This time it was just going to scan my brain constantly in real time and present the continuously shifting contents to the Shrouded.

We tried the drones first, broadcasting at them. But they weren't really set up to react to this kind of thing, except as an exercise in targeting, so we tried through a wired connection that a particularly bold Safeguarding engineer ran to one. Nothing happened. The drones, whose menacing shapes had been hanging over us all this time, turned out to be completely dead, abandoned by the Shrouded.

That was fine, we had a backup plan. We knew the Shrouded were in the ship and we could patch the pod's signal into the *Garveneer*'s comms. We'd just play them my brain's greatest hits as I lay there in the couch, trying to think nice things. Not really much more than the *We are here!* of our simple signal to them, but in this case with a far more detailed model of *we*. Ste Etienne established the link. It was probably the simplest part of the entire operation.

It didn't work either.

Oh, the Pharma read my brain, sure enough. But we'd somehow assumed the Shrouded would recognize this signal was for them. Our olive branch. Our white flag. Our treatise on the nature and existence of that mythical creature, humanity. Except, of course, the Shrouded were shutting down every attempt to link to anything, like someone hitting rats with a stick, and that included my oh-so-clever plan, which we'd just spent far too much of our dwindling time and materiel on.

When they finally called me out – I'd been down in the pod *thinking* as hard as I could – I almost burst into tears. They had given up on us and that sporadic alarm became a mocking thing. A derisive hooting at our imminent demise.

It was definitely colder in here.

'It was a good idea,' Ste Etienne said defensively, against the grumbling. 'It can still work. There's a way.'

Nobody could see the way, except her and, after a moment, me. But I pretended I couldn't. And then, when she explained it to everyone at the top of her voice, I pretended it wouldn't work and it was far too risky and *no*, basically. Just no.

Everyone looked to Chief Director Advent. To say he didn't like the idea was to exhaust the concept of understatement, but then there wasn't much that was likeable going on aboard the *Garveneer* right then.

'Make it happen,' he said. The death sentence.

Ste Etienne sketched out her plan, which would involve just an adaptation of existing tech, stuff we basically had in the workbay that could be repurposed. Transmitters, really – the strongest ones we could muster. We put them where the guns had been. The work took a surprisingly short time, in which I managed to snag Jerennian and make some hard demands and desperate plans with him. Because nobody needed me for the engineering, they'd just need me later. My usual brief, to be the go-between who smoothed things over and got things working. Who *connected*.

When they had it installed, and then tested its deployment, the assembly of aerials looked like the sort of thing a mad conspiracist would wear on their head to keep out the alien mind control rays. Except in this case it was almost exactly the opposite. I stepped down into the pod and linked to the pharmacopoeia again, and they tried to work out whether my brain activity was being broadcast properly. The answer was, 'Maybe but we don't really have the tools to know.' Very reassuring.

I came out once more. My hands were shaking. If everyone thought this plan was a bad idea, then you could put me right at the head of that line.

We stood back from the pod, looking to Chief Director Advent. In his face was the knowledge that a whole community of humans had already set off for the world he had been tasked to prepare as their safe destination. It wasn't just that the Concern was relying on him for its further profit and resources, there were thousands of lives on the line. He was, I was surprised to discover, a better man than he should have been.

I waited for him to tell me I didn't need to do this. I *wanted* him to tell me I didn't need to. Because I'd rather have flushed myself out of an airlock, honestly. But I just couldn't see any other way to break the deadlock between us and the Shrouded.

I did not, under any circumstances, want to go back to the surface of Shroud again. My time down there had been the most traumatic days of my life. Only Ste Etienne's company had made it tolerable. And Ste Etienne didn't want under any circumstances to go back, with or without the dubious benefit of my presence, but everyone else's survival got the casting vote. She and I. Just like old times.

Advent gave the nod. When I didn't initially move, Ste Etienne clapped me on the arm. 'You ready?'

No, of course I wasn't. How could anyone be ready for this? But I nodded and found a smile for her, and made damn sure I got into the pod ahead of her.

She was clambering up the side right after me. The two survivors, the humans who'd beaten Shroud. Except nobody beat Shroud. They'd come up here and beaten the crap out of *us*.

I paused at the top hatch and looked down, catching Jerennian's eye. He nodded and kicked off towards the pod.

I was inside the pod, but they had the cameras and the screens on, to test they were working, so I saw it all. Jerennian grabbing Ste Etienne by the waist, then pushing away from the pod's hull, hauling her kicking body away.

I closed the hatch and sealed it. Advent wanted to know what the hell, basically, and I told him that having two people in here would confuse things. Muddy the waters. This was probably a lie, and I certainly had no hard data to back it up. But if I was going to do any good thing in my life, then it would be to spare Ste Etienne from having to go to Shroud again.

She fought. I saw her practically kneecap Jerennian in her attempts to break free. Her grimy face, smeared with grease, dust and the last of other people's blood, had new clean tracks down it. But then Big Mike wrestled her out into the next section and she was taken out of my sight. Out of harm's way.

They had to evacuate the workbay then, of course. Otherwise I'd be going to Shroud with way more company than I wanted. Everyone left except one figure – I think it was FenJuan. Suited up and crouched by the big manual release lever. We'd found we couldn't actually *evacuate* the workbay, in the sense of take the air out of it in a controlled manner, because those systems weren't working. So I anchored the pod's feet to the deck with magnets and FenJuan had about a dozen straps and cords securing them to the wall. The hatch wasn't supposed to open up with a bay full of air, for obvious reasons, but the safeties were just as dead as everything else.

I thought we'd both be torn loose as the air rushed out. The jaws of the workbay opened, and it felt like a giant was trying to cough us violently from its throat. I saw FenJuan thrash about, and one leg visibly break within their suit, but they were still moving when the air had gone. Even gave me a jerky wave. The suits had their own pharmas and presumably they'd just emptied its painkillers in one go.

Then, before I could have second thoughts or freeze up, I released the feet and set the pod gliding smoothly from the workbay's confines, towards the yawning darkness below. Not even the healthy dark of space, but of Shroud.

Compared to the first time, my exit from the *Garveneer* and entry into Shroud's atmosphere was the most orderly thing imaginable. On account of nothing exploding at the time, mostly. I kept my hands off the controls and let the automatics do their work. I also prayed the Shrouded wouldn't decide to interfere at some critical moment.

I fell into Shroud's gravity well. The chute opened and I felt air resistance and the moon's pull fight over my soul. I had the comms on, so as I dropped towards the surface, I fell into the roar of the world's busy radiosphere, desperately fishing for the distinctive signals of the Shrouded themselves. There were a thousand things that could happen to me that wouldn't even involve them. I'd seen plenty of predators for whom the pod gave off a false signal as a tasty snack. There were angrily territorial monsters and there were volcanoes and there were seas.

As I neared the surface, after that long fall, I started to receive their signals. I didn't believe in being that lucky with my landing spot, and so guessed they'd probably known I was coming. Word sent down from the ship above, registering the pod's launch. Then a spot of mathematics to work out where I would end up. Maths that most human systems would struggle with, given wind variables and all the rest of it, but which I reckoned the Shrouded could just about do in their sleep, now. I was counting on it.

I activated my transmitter. The pharmacopoeia, reading the activity of my brain instant to instant, and then projecting it into the noise-heavy radiosphere. The measure of a working human mind was being written in a dance of electromagnetic activity – the only medium that would mean anything to the Shrouded. A model for them to study, then interact with. Because the functions of the pharmacopoeia meant that the link between model and brain was entirely two-way.

I felt the lurch of landing. My lamps showed me a biome I'd never seen before, the pod wedged between tall, forked stalks bristling with mobile fronds. There was movement below, though. Familiar, deliberate movement. My old friends, who had travelled with Ste Etienne and me for so long.

I thought of all we'd shared, and the transmitters spoke those thoughts out into the world. Then I waited to see if they'd be heard.

6.10 DARKNESS

Your thoughts. I can hear the *Unh unh unh* within them, but only as a part of a swirling whole. I had guessed the insides of your metal bodies were complex, but this . . .

From a thing of brute grunts and hollow spaces, this is too much, and moving too swiftly for me to appreciate it where you have landed. So I carry you into my nearest home, to a buried chamber insulated against the static of the world. Connected to the great web of lines I have drawn, the links I am built from. And there I marvel at what you have sent me.

A message. An explanation perhaps. I draw on more and more of myself to examine you, just as I did with your systems, until the sheer bristling inquisitiveness of my perspective can see you from enough sides to grasp what you are showing me.

Instructions, are they? To construct something that will let me find you. As with your flying machines, and your orders to your killing bodies, it is something of yours that I can copy and adapt and grow stronger through. A last idea of yours, a final gift. A fascinating puzzle.

I watch as your patterns shiver and flicker, a wealth of signal and connection that I can map, copy and analyse, create models and mirrors of, and *become*.

Oddities flicker within me, as I copy segments. As though I briefly have hold of memories I cannot understand. Alien moments in time. Indecipherable ways of feeling and being. But a piecemeal borrowing achieves nothing. What you are showing me exists only in the aggregate, not in its segments. I can understand that.

And then the true revelation comes. As I explore the connections within your body, I discover that the external show of your thoughts is only in itself a model. Linked to it, through channels of signal that I painstakingly map and follow, is the original. A net of interactions that you have so obligingly expanded and laid out for me. I do not even *need* to recreate a model of this multifaceted thing. I have yours to work from.

I grasp, then, why you have sent me this tool. Despite your murderous intent, despite my frustrated response, there is hope for more between us.

It will be the greatest challenge you have set me, to touch – ever so lightly – this construct of data. This kernel of complexity within your body. This *you*.

The strangest of all concepts. That there can be a *you* in a universe where, before, there was only a *me*, even if that *me* was divided into all those Otherlike fragments.

But you have shown me the inner you. You have gnawed at the barriers between us until I can touch you. Until what is *I* and what is *you* are not so different, at the fringes of us both, where the signals meet and mesh and cannot be distinguished.

With the most extraordinary delicacy and care, I touch your thoughts with mine.

EPILOGUE

Mai Ste Etienne is technically under guard, after giving Mikhail Jerennian a black eye and then booting him in the crotch. By which time the workbay had been sealed off and no amount of threats and screaming could get her in.

Ducas FenJuan has been recovered and given medical attention. In considerable pain, despite the drugs their suit dosed them with. Thankfully, zero-gravity is a boon for bone breaks in the short term, though they'll need a whole suite of drugs to make sure everything heals properly.

After that, there is nothing but the waiting. Rigging up heaters, keeping the air flowing, retrieving rations. The entire remaining crew of the *Garveneer*, including those formerly shelved as a waste of resources, sit and consume the last of the ship's reserves, and wait by the comms for an open channel.

Advent has retreated into himself. Taking responsibility in the least useful way, lost to self-recrimination. Terwhin Umbar takes over, keeping discipline and distributing meds. Releasing Ste Etienne on a pledge of good behaviour, holding the engineer's gaze a moment.

'It should have been me. It was even my idea. To go down there.'

Umbar's look is unsympathetic. 'Welcome to how it feels to be in charge.' Who knows how many people have died on her watch, in the service of Technical Oversight's efficiency drives? It happens. Everyone knows that.

Then Umbar maybe sees something in Ste Etienne's face, or feels some after-effect of their brief and pragmatic liaison from before the death of Special Projects. 'She's a hero,' she tells her. 'Someone had to be.' She probably adds something more, some platitude about Juna Ceelander being cited in Concern dispatches, or some other empty honour. But then the channel crackles to life and everyone goes silent. The hum of the air filters is the loudest thing in the universe.

The first sound is dreadful. Just three hissing noises, in a pattern everyone has become horribly familiar with. Whatever is behind it is surely nothing human. This isn't what they've been waiting for after all. But then comes a voice. A woman's staticky voice, carried from the monstrous gravity well beneath, up the elevator cable, and then broadcast across the void to the *Garveneer*.

'Hello?' it says. *'Hello?'*

A few gasps, a sob. Ste Etienne practically lunges forward, all promises of good behaviour forgotten. 'Juna!' she shouts. 'Juna, it's Mai!'

'Hello?' says the voice, and then *'Mai?'* No clue as to whether it's meant or just mimicked, but proof they're being heard.

'Juna, talk to us.' Nobody's wresting control of the conversation from Mai Ste Etienne right then. It'd be more than their bodily integrity would be worth.

A long pause. The hissing, stuttering void of distance between speakers. Physical and perhaps otherwise.

Then it begins, that familiar voice of a woman lost to the worst of all possible worlds.

'Let me put this,' she says, 'into words you can understand . . .'

About the Author

Adrian Tchaikovsky was born in Woodhall Spa, Lincolnshire, has practised law and now writes full time. He's also studied stage-fighting, perpetrated amateur dramatics and has a keen interest in entomology and tabletop games.

Adrian is the author of the critically acclaimed Shadows of the Apt series, the Echoes of the Fall series and other novels, novellas and short stories. *Children of Time* won the prestigious Arthur C. Clarke Award, and *Children of Ruin* and *Shards of Earth* both won the British Science Fiction Award for Best Novel. *And Put Away Childish Things* won the British Fantasy Award for Shorter Fiction, and *The Tiger and the Wolf* won the British Fantasy Award for Best Fantasy Novel. His standalone novel *The Doors of Eden* won the Sidewise Award for Alternate History.

You can find Adrian on Bluesky @aptshadow or at adriantchaikovsky.com